KANZI

The Ape at the Brink of the Human Mind

Sue Savage-Rumbaugh

Roger Lewin

John Wiley & Sons, Inc.

New York • Chichester • Brisbane • Toronto • Singapore

Copyright © 1994 by Sue Savage-Rumbaugh and Roger Lewin.
Published by John Wiley & Sons, Inc.

Photographs on pp. 62, 141, and 182 © Language Research Center, Georgia State University

Library of Congress Cataloging-in-Publication Data:
Savage-Rumbaugh, E. Sue.
   Kanzi : the ape at the brink of the human mind / Sue Savage-Rumbaugh, Roger Lewin.
      p. cm.
   Includes bibliographical references (p.   ) and index.
   ISBN 0-471-58591-2 (cloth)
   ISBN 0-471-15959-X (paper)
   1. Kanzi (Bonobo)   2. Bonobo—Psychology.   3. Chimpanzees—Psychology.
4. Learning in animals.   5. Animal communication.
I. Lewin, Roger.   II. Title.
QL737.P96S26   1994
599.88'440451—dc20   94-9038

This book is dedicated to my mother and father, who gave their entire lives selflessly to their seven children, with never a thought that any parent might do otherwise. They worried over us, they prayed for us, and they tried tirelessly to set the best example they knew how, too often feeling they had fallen short.

My father taught me to value life and honesty and always to give my best to both. My mother taught me that the qualities of grace and kindness may not come easily, but they must constantly be sought, for they form the nexus of all relationships.

When I first began to study apes, they wondered why. When I first was bitten, they *asked* why. I hope that now—they will know why.

Contents

Preface

THIS book really began the day the first joint of my right index finger was severed by an ape I didn't even see. Up from the bowels of a dimly lit cage she raged and parted me from the first joint of my finger. Was she angry at me? She didn't even know me, nor I her. I had just come to the Oklahoma "Chimp Farm" to learn about the signing apes, the ones that were supposed to talk to you with their hands. Little did I know that most adult chimps living in social groups are not kindly disposed to strangers, viewing them as something of a threat to be dealt with immediately. This was long before Jane Goodall had learned that apes kill members of other groups in the wild.

I had begun to study apes only a few months before this bite, but already, within three days of meeting them, I knew that the rest of my life would be spent studying apes. So like us they are, and yet so distinctly different in some ways. It had not taken long to see that human beings could learn a great deal about themselves and the kinds of creatures they might once have been, by studying apes. How much of the ape was left within us and how much of what we had become resulted from the complex society we had managed to build? I was fascinated with this question and knew that many of the keys to its answer lay hidden within these animals. It never occurred to me to stop studying them because I was bitten, either then or on the many occasions that followed. Each time I learned more about what can prompt aggression and grew grateful for the insights, if not for the injuries.

With something of a passion, I set about to find out how

apes become apes. Does it just happen naturally or do they need to be taught how the world works just as human children need to be taught? This book is about one ape out of the eleven that I have studied. This ape, a bonobo named Kanzi, began to learn language on his own, without drills or lessons.

The collaboration that resulted in the telling of Kanzi's story began late one afternoon with a phone call from Roger Lewin, not long after Roger had come to the Language Research Center to see Kanzi for himself. Roger greatly enjoyed his visit, but I was feeling discouraged because in response to everything I had asked Kanzi to do, he either refused, appeared to have no clue as to what I was asking, or did exactly the opposite to what I had asked. Kanzi didn't really like being watched by strangers. When people came to "see Kanzi for themselves," he often tried to determine what sort of a reaction he could get from them by startling them at unsuspecting moments. Kanzi does not like strangers in his space because he cannot understand why they want to come and stare at him. I knew that Roger was thinking of writing a report on the work at the Language Research Center for the *New Scientist,* and given Kanzi's lackluster demonstration, I feared the headline might read, "Brilliant Ape Experiences Memory Lapse."

Having one's scientific findings painstakingly collected over decades thrown into question by the "attitude" of an ape toward visitors has been something that I have had to accept as a consequence of daring to study what is called "ape language." Actually, I never intended to study "ape language" at all. I started out in the late 1960s as a "behaviorist" hoping to discover "principles whereby behavior could be controlled" but came to conclude that "control" was not only an elusive goal, but a dehumanizing one as well. I shifted my focus to attempting to understand the development of the mind in very young children, as it seemed that if this were elucidated, perhaps our society could use such knowledge to build better individuals from the start.

Sometimes the problems of our society are blamed on poverty, other times they are blamed on lack of crime control, and on other occasions they are said to arise from lack of tolerance. We do not yet understand why one individual born into an unstable, lower socio-economic family becomes a president and another becomes a criminal. Is it really just "bad genes"? Or could there be experiences that would enable any child to reach inside himself or herself and pull out the best instead of the worst, as both come packed into each human being?

It is not possible to rear human children experimentally in such a way as to look for answers to this question. Nor can one "watch and see" as children are raised. Twenty to forty years are required to follow children from birth to adulthood. Even when psychologists are prepared to put their careers on hold for this length of time and even if they are able to obtain money for such work, they cannot really watch what happens inside the family. They find themselves limited to interviewing people briefly, giving "standardized assessment tests," and perhaps videotaping some standardized interactions. But since we don't really know why one person succeeds and another does not, we don't know what to "test for." Our tests measure what is easy to measure, whether or not it is relevant, and at the end of such studies, we are still left only with "measurements." We must search to correlate these in a quest to determine which "variables" are related to success, assuming that the recipe for making a person may lie, for example, in the degree of correlation between income and maternal time in the home.

No one in modern societies watches what happens as kids grow up except their parents, and often they do not. In apes, as in "primitive" societies, everyone watches the kids grow up; indeed, everyone is responsible for everyone else.

Over the past two decades I have had the opportunity to watch some apes grow up—apes of different species and very different backgrounds. One thing stands out among a panoply of events: Rearing experiences make a difference. Exposure to people and language does not turn an ape into a human being, but it does result in an ape that can remove itself much further from the exigencies of the moment and reflect in greater depth on the possible consequences of its potential various actions. Such an ape can also understand the intentions of others as expressed through language, though the nonlinguistic expression of intent must match the linguistic one or the words will be ignored.

The first two years of an ape's life are something of a magical time. During this period, if exposed to brightly colored geometric symbols, apes learn to tell them apart as easily as if they were looking at different kinds of food. If exposed to human speech, they become responsive to the phonemes and the morphemes so that spoken language no longer sounds like a string of noises. If they watch television, they come to see the patterns on the screen as representations of other people and other apes in different places, rather than just flickering images. A sense of imagination and narrative begins to emerge, so that they become as interested in "TV" stories to which they can relate as are human children. They grow interested because they can "make sense" out of what they see and out of many of the words they hear. They especially like to watch interactions—interactions between apes as well as interactions between apes and humans. Themes of danger and danger resolved rivet their attention.

When they do not have these experiences, they encounter great difficulty in distinguishing different geometric symbols, in telling sounds from words, and in following an imaginative narrative depicted on a television screen. Often, even extensive training attempts cannot compensate for the lack of such early exposure.

What do observations such as these, and apes like Kanzi, tell us about ourselves, our society, and the early experiences of our own children? The answers to these questions are elusive. Certainly people who have helped rear Kanzi elect to treat their own children very differently with regard to language competency. They recognize that language awareness and comprehension emerge long before a child can say anything, and that the child will demonstrate this awareness if given a chance. Children of such parents recognize symbols and use them to communicate months before they are able to produce intelligible speech, suggesting that the linguistic capacity of the child during the period prior to the onset of speech may be seriously underrated.

But having gained these insights was one thing, convincing

others was yet another. After Roger's visit, I tried to tell myself that it had not gone as poorly as I feared. I had shown Roger videotapes of Kanzi in a better mood, and tried to clarify what Kanzi was capable of doing. I knew, however, that videotapes were sometimes suspect of being selected to eliminate Kanzi's "mistakes," and people often cared only about what they could see for themselves. Because what Kanzi elected to show most strangers was his skill in intimidating them, I recognized ape language had a credibility problem.

What I did not realize was the depth in which Roger had already investigated the work of the Language Research Center. He had taken the time that was needed to study the scientific reports from the center that now spanned two decades and more than three hundred publications, and to watch other tapes that detailed what Kanzi and other apes were capable of doing. He recognized that scientific findings often unfold in an organic manner and that one can neither fake real findings nor show them off. They emerge because one keeps searching in every possible manner for explanations to complex phenomena. The understandings that arise are of a "whole cloth" and function to help us see ourselves from new perspectives. They could not have been "shown off" in a brief visit even if Kanzi had elected to cooperate.

Roger called me later to thank me for the opportunity to visit. I was happily surprised. He told me that he had been deeply impressed with all that he had seen and urged me to write a book that would convey much of what I had been compelled to leave out in scientific research reports. I agreed that it was important to attempt, but saw no time window. Caring for all of the apes, while attempting to accomplish research, was already more than a full-time job. I asked if he would help. He agreed. I was surprised yet again, and much pleased. So enmeshed had I become in the work that things to which I no longer gave a second thought took others by surprise and required explanation. I knew I needed someone to work on the story with me; it was too all-encompassing to tackle alone. And so we set about the task together. The book that follows, though written in the first person, represents a joint effort to portray a story we have both come to understand and appreciate far better by working together. Although the firsthand experiences were mine, in sharing them with Roger they became, in a sense, our joint experiences, interpreted through four eyes instead of two. Roger never walked in the woods with Kanzi, but the story quickly became his as well as mine because he understood it, and then we worked to make it ours. I hope we have succeeded.

Sue Savage-Rumbaugh

Acknowledgments

MANY people have played a role in this story over the years, each with his or her own attraction to and insight into the mind of the ape. Four stand out very clearly as having been an integral part of all that has happened and all that continues to unfold each day. Duane Rumbaugh has provided constant drive, constant encouragement, and constant support. He has made me believe that what is being learned is worth the effort and will, someday, be of value. Rose Sevcik has been with me and with all the apes through every problem and every crisis, and there have been many. Without her aid and steadfast courage, there are times that I could not have gone forward. Her belief in what Kanzi can do and who he is has made all the difference. Liz Pugh has been there as well. She has gotten me out of cages when Austin and Panzee locked me in, and she has shown Kanzi and Panbanisha they can do a thousand things they thought were impossible. She has raised a child as well as four apes, and no one knows better than she how close we really are. Mary Ann Romski has made everything we have done and learned real, as she has worked more responsibly than anyone will ever know to translate the best of what we have learned into real and pragmatic help for human children. But her efforts have not stopped at help, they have gone far into the scientific documentation of why what we have learned really works.

Brent Swenson, or "Dr. Brent" as we refer to him, has kept Kanzi and many other of our apes alive on more than one occasion. He has seen them through air sac infections, severe allergy attacks, heart defects, herpes, strep, flu, injuries, pneumonia, abscessed teeth, and many other ailments that befall man and ape alike. He has done it with far less "information" than most medical doctors have at their disposal, for the apes do not cooperate so willingly with regard to standard medical inspections. He has also done it with a patience that is extraordinary.

Without his wealth of knowledge, his keen eye, and his sensitivity to all that apes do when they are ill, we never could have succeeded in learning about Kanzi. The first bonobo that was studied in captivity by Robert Yerkes died around four to five years of age. By contrast, the advent of modern medicine, when practiced by a person of great sensitivity, skill, and special concern for bonobos, has made possible research that surely would have been terminated in an untimely manner in an earlier day.

While I was working on the book, many people provided advice, encouragement, and support. Shelly Williams made certain that the lab continued to run effectively and that Kanzi and the other apes were not only happy but involved with life and with research in a creative and positive way. Iain Davidson, Patricia Greenfield, Duane Rumbaugh, Mary Ann Romski, Rose Sevcik, and Nick Toth all participated far beyond the ordinary in helping with the manuscript itself, in its detail and its story. Some of the excellent photos have been provided by Nick Nichols, taken with insight and intensity out of concern for the plight of apes everywhere.

Solid encouragement that the story being told was worthwhile and that its philosophical implications could be successfully conveyed has been provided by Stuart Shanker.

Many people continue to try to make certain each day that the apes in our laboratory receive every opportunity to express themselves both to human beings and to other apes, and to live lives full of social interchange and self-dignity. To the people who give themselves so selflessly to these apes and who are trying hard to see through the rough exterior that apes project to the depths of the gentle souls that rest underneath, I am deeply grateful. Thank you Adrea Clay, Angela Fox, John Kelly, Linda McGarra, Julie Meitz, Jeannine Murphy, Skip Haig, Jane Patton, Phillip Shaw, Dan Rice, Ryan Sheldon, and David Washburn.

The apes at the Language Research Center, with the exception of Matata, have all been born either at the Yerkes Primate Center or at the Language Research Center of Georgia State University. Without the trust and confidence of the Yerkes Center, which permitted us to work with these apes, this story could never have begun to unfold. Without the unflagging support of Georgia State University, which has built our facility, provided us with a forest, and stood by the value of our work through attacks from animal rightists as well as those who believe apes have no business learning language, the story of these apes would have vanished long ago. And without the peer support provided by panel after panel of site visitors from the National Institute of Child Health and Human Development, we would never have had the funds to make any inroads into the area of ignorance that surrounded our understanding of the intellectual capacities of apes.

And without Lana, Sherman, Austin, Mercury, Panzee, Panbanisha, Kanzi, Neema, Matata, and Tamuli—well, without them, my life would have been stripped of meaning. They have shown me what it means, and what it takes, truly to be a human being. I thank them for the countless lessons they have given me on becoming a window through which others may shine.

— 1 —

On a Beach in Portugal

THREADING my way along the sandy path toward the ocean shore, I sought out the rhythmic sound of shifting surf. The faint light of predawn arrived and I could see the rocky coastline ahead, then the silhouette of the distant mountains behind which the sun would soon rise. I was near the small coastal village of Cascais, Portugal, attending a meeting organized by the Wenner-Gren Foundation, a group lengendary in anthropological circles.

Scientists invited to Wenner-Gren conferences are kept away from the rest of the world and encouraged to examine each other's views in small and intense conferences. Until recently, such conferences had always taken place at "The Castle," in Burg Wartenstein, Austria. But times being as they were, even the Wenner-Gren Foundation could no longer afford the luxury of a castle and its attendant staff, solely for the purpose of getting scientists to talk meaningfully to one another.

Having forfeited its beloved castle, the foundation located a small hotel near Cascais, not least because the hotel, though relatively new, was constructed as a castle. Ancient stone had been formed into thick heavy walls with arched windows, enclosing restful courtyards. This modern bastion was poised

1

high on a bluff overlooking the Atlantic. The setting rivaled that of the original castle—especially in its isolation.

Early morning is the time I try to bring order to the flood of thoughts running through my mind upon waking. Once the bustle of the day begins, I must constantly be prepared to respond to others and usually have little or no time to reflect. Thus I find it best to try to seek a bit of solitude before the day arrives, when I can. Walking along the beach at Cascais, I mused over the discussions of the past few days. Bill Calvin, a neurobiologist at the University of Washington, Seattle, had been talking about the extraordinary accuracy and power with which humans can throw. Chimpanzees and gorillas can throw, too, as visitors to zoos sometimes discover, to their chagrin. Apes do not enjoy being stared at and frequently throw things at visitors in an attempt to make them leave.

Calvin was among the most thoughtful of the scientists gathered about the round table overlooking the ocean where we spent most of our daylight hours. Unlike others who could not wait to disagree, Calvin took in information and permitted it to "interact" with the vast pool of knowledge he already had stored. Only when some new insight arrived as a result of this process did Calvin wade into the constant fray that was taking place around the table. He, more than the rest of us, knew how brains worked, and he was very good at letting his own brain have the room it needed to do its job. It took a while to realize this, but when I did, I made certain to listen carefully whenever Bill Calvin decided to speak.

The development of throwing, Calvin had pointed out, was clearly important during man's evolution from an apelike ancestor. In particular, the accurate hurtling of stones became a valuable means of hunting and self-defense against predators. Another scientist, Nick Toth, was also interested in throwing, but for a different reason. Nick, unlike the rest of us, actually knew how to make the stone tools that our prehuman ancestors had utilized.

Nick was not a typical scientist. I recognized this right away when he sat down with his briefcase and began pulling fist-sized rocks out of it. He then casually mentioned that the most monumental decision of his life had been whether to be a rock musician or an anthropologist. Interesting dilemma, I thought, the only commonality being that both professions focused on rock.

The previous day, Toth had riveted the group's attentions with his display of stones and demonstrations of how rocks can become tools. He explained the physics of conchoidal fracture, by which good, sharp flakes can be made, and he challenged us to accompany him to the beach to try to make the "crude" stone tools of our two-million-year-old ancestors, *Homo erectus*. That afternoon I gained a newfound respect for the feats of my "prehuman" ancestors. No longer did I deem it appropriate to apply the word *crude* to their tools.

It was my first attempt to emulate a Paleolithic stone knapper, and I did not find it an easy task. Neither I, nor most of the other "educated scientists," could coax even a single flake from the pebbles on the beach during our first half hour of trying. We even resorted to placing one stone on the ground and slamming another against it, but to no avail. Finally, instead of just watching Nick, I began to look closely at what he was doing. Why did the stones break so easily when he struck them together with such little force, while they just made a loud "thud" when I slammed them together as hard as I could?

I finally recognized that Nick was not really hitting rocks together; instead, he was throwing the rock in his right hand against the edge of the rock in his left hand, letting the force of the controlled throw knock off the flake. The "hammer rock" never really left his right hand, but it was nonetheless thrown, as a missile, against the "core," or the rock held in place in his left hand. What had I been doing? Just slamming two rocks together as though I were clapping my hands with rocks in between.

Once I realized how Nick was actually flaking stone, I grasped the profound similarity between the activities of throwing and stone knapping. In each activity, you must be able to snap the wrist rapidly forward at just the right moment during the downward motion of the forearm. This wrist-cocking action produces great force, either for achieving distance in throwing or for knocking a flake off a core pebble. I also knew that the wrist anatomy of African apes prevented them from making this kind of movement. Chimpanzees' wrists stiffened as they became adept knuckle walkers. They cannot bend their hands backward at the wrist as we can, but they can put weight on their hands for long periods without injury to their wrists.

In addition to the force of delivery produced by the wrist snapping backward during throwing, it is also important to deliver your blow to the core accurately if you need to detach a flake. Several of us had bruised fingers after the afternoon's stone-knapping excursion, suggesting that, accurate though we might be as a species, as individuals we needed practice.

Bill Calvin was likely correct in his suggestion that throwing ability had been selected for in the course of human evolution. But accurate throwers also had the potential skills to make stone tools. Could throwing as a defensive device have paved the way for the deliberate construction of stone tools? Certainly to use throwing as a means of defense, it would help to be a biped and to have a wrist that could snap backward to launch the missile. Evolution often finds a way to make interesting and unexpected connections between anatomy and behavior. For instance, bird feathers might have evolved initially as a means of insulation, but they are essential for the behavior we call flight.

Our conference, which was called "Tools, Language, and Intelligence: Evolutionary Implications," was searching for evolutionary links between language, tools, and anatomy that could have led to the emergence of the bipedal, large-brained, technological creature that is *Homo sapiens*. As a psychobiologist, I found the neurobiology of stone throwing and the skills of stone knapping entirely new, and I was intrigued. As I walked on the beach, I attempted to integrate these ideas with my own knowledge of how apes came to understand and effectively employ both tools and language. Suddenly, I noticed the outline of a dog standing on a slight rise, about ten yards ahead of me, motionless.

Being accustomed to encountering abandoned dogs in the

forest around my laboratory, I didn't take much notice initially. Then another appeared to my right, and then another. Soon there were at least half a dozen, and they began to emit a low, ominous growl as they slowly formed a tight circle about me with their heads lowered and all eyes focused directly on me. I remembered having heard about packs of feral dogs in the region, wild and isolated as it was, and as the circle began to tighten, I came to the distinct conviction that I was being hunted. I looked back toward the castle and realized that no one else was either up or outdoors, and through the thick walls no one could even hear a scream. Thus I did not bother.

My memory flashed back to when I was five years old and a single dog had approached me growling in exactly the same way. I recalled standing my ground as long as I could and then turning and running in terror. As soon as I began to run, the dog attacked. I was not about to run now. But what else could I do? A vision of a naked prehominid standing on a similar windy bank, encircled by predators of some odd form, flashed in my mind. This prehominid female leaned down, picked up a rock, and threw it with great force. I did the same.

Luckily, I scored a direct hit on the buttocks of one of the dogs, and the recipient yelped. Hearing this, the others began to back off. I followed them, still throwing, to show that I intended to take the offensive while I had the upper hand. Finally the pack turned and fled. I hurried back to the cliff path that led to the safety of the hotel, realizing with a new appreciation that Bill Calvin certainly had a point: A bipedal creature devoid of natural weapons can do much to ensure its survival with an accurate, powerful throw. From that moment on, speculations regarding the "survival value" of various skills supposedly possessed by our ancestors took on a more vivid realism for me. These weren't just academic musings we were engaged in. We were touching upon the real, honest-to-goodness life events that determined whether or not a given individual lived to be a mother or a grandmother and thus whether or not there were beings to carry on the human form.

For the two decades I have known and studied chimpanzees, I have been attempting to discern the degree to which they can think and communicate as we do. For many reasons—some justified, some not—this work has been among the most controversial areas of academic endeavor. The initial efforts of ape-language researchers, in the 1960s and early 1970s, were hurriedly greeted with acclaim by the popular and scientific press alike. Newspapers and scientific journals declared the same message: Apes can use symbols in a way that echoes the structure of human language, albeit in a modest manner. Symbol use heralded an insight into the evolutionary connection between ape and human brains, it was said. The symbols were not in the form of spoken words, of course, but were produced variously as hand gestures from American Sign Language, as colored, plastic shapes, and as arbitrary lexigrams on a computer keyboard.

But in the late 1970s and early 1980s, this fascination turned to cynicism. Linguists asserted that apes were merely mimicking their caretakers and that they displayed no language-like capacity at all. At best, critics viewed the research as flawed; at worst, they suggested it was an example of either egregious self-deception or outright fraud. Multiple reasons lay behind this dramatic shift of opinion, some scientific, some sociological, and I will explore some of these reasons later.

These were trying times, as all scientists working in the field were peremptorily accused and convicted of "poor science," whether guilty or not. Most linguists and psychologists simply wanted to forget apes and move ahead with what they viewed as the "proper study of man"—generally typified by the analysis of the problem-solving strategies of freshmen students. These researchers did not want to be bothered with questions about whether or not an ape knew what it was saying.

From my earliest exposure to apes, I recognized there would be considerable difficulty in determining whether or not they employed words with intent and meaning in the same way that we did. I also believed that the pioneers in the field of ape language had not fully grappled with the issue of word meaning. Because the intellectual excitement at uncovering language-like faculties in apes had been intense, the pressure to overinterpret was immense. It seemed that anything an ape signed was accepted, so much so that ape-language research came to be labeled a "nonscience"—to some, it even became synonymous with nonsense.

I persisted in trying to come to grips with the essence of ape language only because, having been granted the opportunity to come to know apes better than I knew most people, I had no doubt they had a great deal to tell us about who we humans were, where we came from, and where the biological limits or constraints upon our species were to be found. I also believed that many behavioral scientists were wasting their efforts in attempting to develop conceptual models of the different "logical functions" of animal and human minds by comparing rats, pigeons, and college freshmen. Apes share 99 percent of our DNA and have a developmental period clearly as profound and plastic as our own. Why were we not trying harder to understand what apes could and could not do, what sort of creatures they were and were not?

From 1975 to 1990, I searched for scientifically credible ways to approach these fundamental questions about apes and their intellectual and emotional capacities. By 1990, the year of the Wenner-Gren conference, I knew that at least some of this work was reaching an audience, or I would not have been invited to the conference. Perhaps there I could start to explain what I had learned. At least in such a conference, I assumed, other scientists would have to listen whether they wished to or not. Unlike most meetings, they couldn't elect simply not to attend the paper on ape language. I would at least have a chance to begin to tell my story or, more accurately, Kanzi's story.

Like all invitees, I had written a scientific paper for the event, describing my research and conclusions. The papers, after revision, were to be published together as a conference volume. But even though I could describe on paper, with proper scientific documentation, what Kanzi did, I knew that I needed to show people images of Kanzi as a living, breathing, thinking

being. My words and numbers were but the pale bits and fragments we call data, data that was dwarfed by the presence and
power of Kanzi himself.

Even reducing Kanzi's 150-pound frame to a small, two-
dimensional television screen seemed to do him an injustice.
Still, I knew his ever alert and questioning intellect would convey the presence of mind within the apelike form. Images of
Kanzi never fail to have a message; they announce, "I am here, I
am Kanzi, I am an ape, I am thinking about my actions, and I
am listening to what you have to say to me." Never could any
numbers that I might compile convey this message to other scientists, but I knew Kanzi himself could do so. Thus I prepared
for Cascais a forty-five-minute video that illustrated the kinds of
things Kanzi could do.

Nevertheless, I felt hesitant when I arrived in Cascais—perhaps because of the countless past occasions when people had
refused to listen to new data and new interpretations, simply
because it was ape-language research. Shortly after I arrived at
the castle, I met Sydel Silverman, president of the Wenner-Gren
Foundation, who invited me to lunch. She showed a great interest in apes and what it was like to work directly with them.
Consequently, I used this opportunity to tell her that I wanted
to show a videotape during my presentation, as so many questions were hard to answer with words alone.

She seemed agreeable to the general idea, but upon checking the schedule concluded that I could be permitted only a
five-minute slot for my video of Kanzi—and this would have to
be allotted at the end of the meeting. It seemed that many scientists who had been studying apes in the wild were also going
to be at the meeting and they also wished to present videotapes.
Their tapes would be given priority because the behaviors they
would be showing were natural. If wild apes were allotted several hours of "visual time" and Kanzi only five minutes, it was
clear that the doubts about ape-language work would have little
chance of being challenged.

I was disappointed; in fact, at that point I would have been
happy to leave. But leaving is something you do not do at
Wenner-Gren conferences no matter how much you may want

to. For one thing, they had taken my passport upon arrival at the castle—just to check on things, they assured me.

This meeting was to be different from others in recent memory, Silverman told me, because it had as its focus the "big questions" of anthropology—the questions that lie at the roots of the field but that somehow had been ignored in the past decade. Well, I certainly had been wrestling with some of those "big" questions; didn't anyone want to see visual documentation of what I had learned?

The meeting was organized by Kathleen Gibson, an anthropologist, and Tim Ingold, a sociologist. Together they persuaded the Wenner-Gren Foundation that anthropology was ready for a new look at human evolution. Earlier, at another meeting in Trieste, Italy, Ingold and Gibson had inevitably found themselves discussing the confluence between the most basic of anthropological and sociological issues. Gibson wondered how our past shaped our abilities to perceive and organize our societies, and Ingold wondered how our societies had shaped and organized our perceptions of the world and consequently our history. In essence, they felt that science could no longer avoid the central Big Question: "What is the nature of the human mind, how did it arise, and how does the construction of mind affect mind itself?"

This question has gnawed at our species for millennia; philosophers, laymen, and theologians alike have grappled with it since the dawn of the human mind, and it was to become the subsuming focus of all our minds during that week on the beach. What is the relationship between humans and the rest of the animal world? Is there a smooth biological continuity between our minds and those of other creatures? Or is there a sharp discontinuity, a gulf so qualitatively great as to be unbridgeable?

———

The scientists who attended the Wenner-Gren conference brought with them the bulky baggage of European man's attempts to come to grips with his place in the natural order of things. In the pre-evolutionary world of Western science, the place of *Homo sapiens* in nature was presumed to be revealed by the Great Chain of Being. Along this chain, academic minds arranged organisms from the lowest level of perfection to the highest, with God at the top and humans "a little lower than the angels," according to Alexander Pope. The Great Chain of Being was thought to reflect the divine ordering of entities and to have been set down by the hand of God, for all time.

From our modern perspective, the chain appears neither "great" nor accurate. Not surprisingly, the chain embodied an explicitly racist view of modern humans. "Ascending the line of gradation, we come at last to the white European; who being most removed from the brute creation, may, on that account, be considered as the most beautiful of the human race," said Charles White, a British physician, in 1799.[1] White's clearly self-serving view was commonly held among scholars at the time and persisted well into the twentieth century.

Although European scholars were confident about the exalted place of "white races" in the Great Chain of Being, there was a problem. According to the theory, the chain should be physically continuous, with no gaps. But gaps there were: namely, between minerals and plants, between plants and animals, and, most embarrassingly of all, between humans and animals. That is, there appeared to be no half-animal half-human beast to fill the gap between ourselves and the other creatures of God's planet.

So pressing was the need to fill this gap that in 1758, when Carolus Linnaeus established the basis of zoological classification with his *Systema Naturae,* he postulated the existence of a primitive form of human, *Homo troglodytes,* to fill the chasm. Explorations of Africa and other distant lands were beginning at this time, and tales of apes and primitive tribes filtered back to the scientific establishment. With only the flimsiest information to go on, scholars constructed descriptions of apes that were half-human and half-ape. Scientific illustrations of the time captured these conceptions, revealing more about the mind of the artist and the perceptions of the time than about the anatomy of the creatures they supposedly depicted. Thus was the physical gap between humans and the rest of nature effectively closed, as demanded by the Great Chain of Being.

Despite the physical continuity among living forms that this sleight of mind achieved, it must be noted that Western tradition continued to hold humans distinct. From Plato to Aristotle to the Christian era, man set himself apart from the rest of the living world by virtue of being the single possessor of a "rational soul," as embodied in the gift of speech. Physical continuity between man and other creatures was tolerated, but when it came to the realm of the mind, popular belief of the time held that the Great Chain had been irrevocably *dis*joined by a God who created man in his own image and then gave him dominion over all the beasts.

This conceit of mind-body disunion, passed down through the generations, imbued otherwise scholarly scientific efforts with a peculiarly myopic perspective as regards the mind. Veneration of the human soul became transformed into veneration of the human mind, the human brain, and all the products thereof: language, rational thought, music, art, and culture.

With the appearance of Darwin's *The Origins of the Species* in 1859, this view began to shift a little. It became necessary to view the Great Chain of Being as alterable, not static and set for all time by the hand of a divine creator. Evolutionary theory revealed a dynamic image of historical change, of "descent with modification," as Darwin put it.

To scientists of the day, the evolution of *Homo sapiens* became a virtual inevitability, a manifest destiny of gradation of form and function from ape to man that had to occur on the way to the production of the pinnacle of the natural world, the mind of man. Whether or not half-man half-ape creatures currently existed was irrelevant, for they surely existed in the relatively recent past. It was assumed that ancestral protohumans became more technological and more intelligent through time and that they were predestined to hold sway over all the earth. European man assumed that at some dim point in the past, the human mind had emerged into the light of awareness, forever distancing man from his younger brother, the African ape.

Thus, although evolutionary theory offered a view of the

Great Chain of Being as under continual construction, the theory itself became transformed to support the separatist view that the minds of men and animals were distinctly different. Along with this view traveled the equally self-serving perspective that evolutionary theory explained why some races had made more technological progress than others. For example, it was believed that European man had simply evolved higher in the evolutionary scale; non-European races were thought to be in a state of transition and therefore intellectually inferior.

Thus was evolutionary theory used, as are all scientific theories, to support the biases of the society that gave it birth. Both racism and speciesism were justified. For instance, Roy Chapman Andrews, a leading figure at the American Museum of Natural History, New York, wrote in 1948: "The progress of the different races was unequal. Some developed into masters of the world at an incredible speed. But the Tasmanians, who became extinct about 1870, and the existing Australian aborigines lagged far behind . . . not much advanced beyond the stages of Neanderthal man."[2]

Lest Andrews be perceived as peculiarly pompous, all one needed to do was attend the St. Louis World's Fair of 1904. There, various races of people were on display as an illustration of the evolutionary stages of human beings from primitive man to white European society. According to Phillips Verner Bradford and Harvey Blume, authors of *Ota Benga*, a book about "anthropological fashions," the eminent anthropologist "Chief McGee," the director of the anthropology section, dispatched special agents of the fair to the four corners of the earth to assemble "representatives of all the world's races, ranging from the smallest pygmies to the most gigantic peoples, from the darkest blacks to the dominant whites."[3] Anthropology wanted to start with "the lowest known culture," and work its way up to man's "highest" culture symbolized by the exposition itself. Bradford and Blume write that

> McGee regretted that it was impossible to exhibit examples of "all the world's peoples on the Exposition grounds." The Anthropology Department had to settle for being less definitive than Carl Hegenbeck's Circus, also featured at

the fair, with its "largest representation of an animal paradise ever constructed."[4]

The human pygmies were kept in an "enclosure" and not permitted to roam the compound at will. They were fed very little and clothed not at all, for it was believed that they neither required nor cared for these amenities in their current state. The chimpanzees and monkeys on exhibit were housed with these people, as it was thought that since the pygmies were closer on the evolutionary ladder to such creatures than to Europeans, they could communicate with them more effectively.

According to the *Scala Natura* of the time, such individuals were viewed as genetically inferior, capable only of concrete, not abstract, thought. These people existed only in hunter-gatherer societies, it was said, because they had not evolved sufficiently to comprehend, much less create, the complex systems of laws and monetary exchange that characterized the modern human societies of northern climates.

Even though his theory was employed to justify such disreputable ends, Darwin himself posited no sharp watershed between man and ape, or between races of man. Indeed, Darwin recognized that the greatest potential contribution of evolutionary ideas regarding the *Scala Natura* could be to help man understand himself through the study of the behavior of his living kin, especially his younger brother in evolutionary time—the ape.

In a widely unread but penetrating book, *The Expression of Emotions in Animals and Man*, Darwin made a spectacular first attempt at tackling the study of the continuity of the expression of emotion. He saw most, if not all, of our human emotions as extant, both physiologically and psychologically, in other mammals to some degree, and he attempted to trace the transformation of these expressions of emotion across a number of species.[5] His work in this area was careful and grounded to the greatest possible extent in the available knowledge of muscular function and anatomy.

However, unlike the physical samples that he collected for *The Origin of the Species*, examples of behavior were impossible to collect and contrast, since at that time there was no way to record actions on videotape; even photographs were difficult to make and reproduce. Moreover, travel being what it was at the time, one could not observe most animals under natural circumstances in the field. For these reasons Darwin requested and relied on the accounts of a variety of other parties who had witnessed examples of communication in animals and who recognized similarities between the communicative behaviors of animals and those of man.

Darwin recognized the shortcomings of such accounts and took considerable pains to focus on directly observable behaviors and to seek out reports from multiple parties. When he noted disagreements, he attempted to verify observations for himself. For example, when he found reports of tears in monkeys as an expression of emotion similar to grief, he attempted to observe this phenomenon for himself; he could not verify it.

Whereas Darwin, in 1872, focused on the continuity of the expression of emotion, Georges Romanes set out in 1886 to expand the area of behavioral continuity between man and animal by including mind and intellect. Taking Darwin's work on the continuity of emotion as a starting point, Romanes observed:

> The expression of fear or affection by a dog involves quite as distinctive and complex a series of neuro-muscular actions as does the expression of similar emotions by a human being: and therefore, if the evidence of corresponding mental states is held to be inadequate [for the existence of mind] in the one case, it must in consistency be held similarly inadequate in the other. And likewise, of course, with all other exhibitions of mental life.[6]

Starting with mollusk, then working his way through ants, bees, wasps, termites, spiders, scorpions, fish, reptiles, birds, marsupials, bison, horses, rodents, bats, seals, beavers, elephants, cats, dogs, monkeys, and apes, Romanes took on the much too monumental task of attempting to assess the degree to which each possessed the powers of reason. Like Darwin, he had to rely on the observations of others and, like Darwin, he took pains to attempt to cross-validate these observations. Many were undoubtedly valid.

For example, as foxes became familiar with traps, they could no longer be caught; consequently, trappers were often forced to seek new methods, a sort of natural experiment in a sense, though one that certainly would not pass the shield of an IACUC (Institutional Animal Care and Use Committee) in this day and time.

Romanes reports the account of one trapper who accordingly

> set a kind of trap with which the foxes in that part of the country were not acquainted. This consisted of a loaded gun set upon a stand pointing at the bait, so that when the fox seized the bait he discharged the gun and thus committed suicide. In this arrangement the gun was separated from the bait by a distance of about 30 yards, and the string which connected the trigger with the bait was concealed throughout nearly its whole distance in the snow. The gun-trap thus set was successful in killing one fox, but never in killing a second; for the foxes afterwards adapted either of two devices whereby to secure the bait without injuring themselves. One of the devices was to bite through the string at its exposed part near the trigger, and the other device was to burrow up to the bait through the snow at right angles to the line of fire.[7]

Unfortunately, observations such as these never resulted in further serious scientific study, which was an intellectual tragedy, for it placed the real impact of evolutionary thought outside our collective awareness. There were two reasons for this sad state of affairs. First, some of the sources Romanes depended on proved unreliable. Some reports were made by people unfamiliar with the general habits of the animals they observed, and behaviors that were attributed to reason were later shown not to require causal understanding. Second, the few psychologists who did undertake the experimental study of these issues did not fully grasp the nature of the problem they were studying.

The most famous example of such work was that by Edward Thorndike, who placed a cat in a puzzle box and carefully watched how it learned to unlatch the handle and escape. He found that it was only by accident, not by power of reason, that the cat became able to unhinge the door. He quickly concluded that most other expressions of animal intelligence were similarly the result of "trial and error" learning. In this example, locking a cat in a box is certain to result in a great many attempts to paw at the door, and some of these attempts are likely to be successful without intent. The cat then simply repeats them once their success is demonstrated.

What Thorndike did not do was to study how the cat might have been able to use the knowledge gained accidentally in one situation insightfully in a different, albeit similar situation. He also did not investigate situations in which the animal itself selected its course of action and the time of execution, as was the case in the previous example given for foxes. Surely, even in man, the powers of reason are most evident when we are permitted to elect both the time and course of our action. Even children locked in a room will engage in an apparently random series of actions in an attempt to escape. On the basis of all too little data, in very restricted settings, psychologists in these early days hastily determined that there was no basis for Romanes or Darwin to conclude that animals could think.

Darwin's work on animal emotions was never disputed nor soundly condemned as was that of Romanes on animal intellect. Instead it suffered the more ignoble fate of simply being ignored. People could not realistically maintain that animals did not manifest emotions as easily as they could maintain that they did not manifest powers of reason. Nonetheless, the continuity of emotional experience was denied. Man was said to experience emotion differently from animals, although it was granted that animals and humans displayed some emotions similarly.

The basis for assuming a discontinuity of mind between man and animal lay in the special interrelationship said to exist between language and consciousness. Language, it was maintained, led to consciousness, and consciousness led to the experiencing of emotions (guilt, sorrow, remorse, exultation, hatred, empathy, and so on). Consequently, the emotions on the faces of animals portrayed little more than simple passing sensations, quite different from the true emotions that were assumed to be reserved for man alone. Science even gave the emotions of man and animals different names to emphasize this point. Thus we call a human expression a smile, and an ape expression an "open-mouth bared-teeth grin," noting that as it is not possible to determine how apes feel when they display this expression, it is best to use terms that do not impute feeling to them.

———

Beginning in the 1940s, anthropology leaned toward a new version of the "human uniqueness" story. Influenced by the so-called Neo-Darwinian Synthesis, which emphasized the importance of adaptation as a metric of physical and behavioral change, anthropologists concluded that the origin of the human family involved a major shift in behavior. This shift led to a lifestyle that rapidly became distinctly different from the subsistence strategies practiced by apes and monkeys. Humans, particularly male humans, became hunters. With the emergence of Man as Hunter, and the associated behavioral changes wrought by such a lifestyle, the gulf between man and other creatures on the planet suddenly became a chasm. Hunting led to planning ahead, to home bases, to kinship ties, to tool construction, to ritual and knowledge of the seasons, and to cooperation—all characteristics thought to set the human mind apart from the ape mind. Consciousness, it was determined, awoke in the form of Man the Hunter.

The drawback of this perspective was twofold: First, it eliminated women from the "Great Changes" that defined *Homo* as different from Ape. Presumably, women evolved, to the extent that they did, because of the activities of men. Second, it put in place an irrevocable boundary between man and all other primates. However, as anthropology made the man-animal boundary central to its world view, it also finally rid itself of its racist perspective. As a result, all human races were considered equal.

The gap between humans and the rest of the animal world at once grew wider, both physically and in the realm of mind. Humans began to look like extremely special creatures, with an ever greater discontinuity separating them from the brutes. To make this world view accommodate practical experience, scientists had to belittle the apparent accomplishment of animals, questioning animals' every act in a way they never questioned their own.

Perhaps it is not surprising that scholars felt comfortable treating *Homo sapiens* as special in a scientific context. After all, we do *feel* special. Even Alfred Russel Wallace, the co-inventor of the theory of natural selection, could not bring himself to acknowledge the total impact of what Darwin had said. "I fully accept Mr. Darwin's conclusion as to the essential identity of man's bodily structure with that of the higher mammalia, and his descent from some ancestral form common to man and the anthropoid apes," he wrote in 1889.[8] However, man's intellectual powers and moral sense, "could not have been developed by variation and natural selection alone, and . . . , therefore, some other influence, law, or agency is required to account for them."[9]

Wallace's reasoning was simple. "Natural selection could only have endowed the savage with a brain a little superior to that of an ape, whereas he actually possesses one but very little inferior to that of the average members of our learned societies."[10] In other words, even humans living in "primitive societies" were more intelligent than they had to be. Adaptation by natural selection should have equipped them only for the limited exigencies of a foraging existence.

If this argument sounds like an unfounded justification for the uniqueness of man, compare it to an observation by David Premack, an ape-language researcher at the University of Pennsylvania. "Human language is an embarrassment for evolutionary theory," he wrote in 1985, "because it is vastly more powerful than one can account for in terms of selective fitness." Although Premack does not, like Wallace, appeal to spiritual intervention to account for the special qualities of the human mind, he nonetheless carves up the world of the mind in a similar manner.

Such personal dissections represent what Oxford University evolutionary biologist Richard Dawkins calls "the

Argument from Personal Incredulity." These arguments proceed from the individual's personal and egocentric bias that his or her mind works in a qualitatively different manner from that of animals. Having established this private conclusion, the proponent of this theory brings into play the known and accepted principles of science for support. The trouble with this method is that not only can it be used to differentiate the animal mind from the human mind, but it also can be (and has been) used to differentiate the "criminal" mind from the "noncriminal" mind, the "male" mind from the "female" mind, the "white" mind from the "black" mind, the "Christian" mind from the "Muslim" mind, the "Catholic" mind from the "Protestant" mind, and to make any other distinction that becomes politically expedient.

During the 1950s and 1960s the thrust of anthropology continued to emphasize human uniqueness, a view in which Leslie White, of the University of Michigan, was extremely influential. Referring specifically to language abilities, he said in 1949: "Because human behavior is symbol behavior and since the behavior of infra-human species is non-symbolic, it follows that we can learn nothing about human behavior from observations upon or experiments with lower animals."[11] Indeed, so impressed by our mental superiority was Julian Huxley, grandson of Darwin's champion, Thomas Henry Huxley, that he opined *Homo sapiens* should be removed from the animal kingdom entirely and be allotted a kingdom of its own: the Psychozoan.

Add to this the notion that only humans use tools, only humans are conscious, and only humans elaborate culture, and a firm boundary between humans and other animals is in place. "In one way or another, policing and maintaining that boundary has been a tacit objective of most paleoanthropological model-building since the late 1940s," observes Matt Cartmill, an anthropologist at Duke University.[12]

The first bricks in this well-policed boundary wall were dislodged in the early 1960s with Jane Goodall's observations of tool-use and tool-making in chimpanzees, at Gombe Stream

Reserve, Tanzania. Goodall had the patience systematically to watch chimpanzees instead of making sweeping conclusions about what they could or could not do based on a few contacts. She saw them stripping twigs to use as probes in fishing for termites. Since then, observers have noted many additional kinds of tool-using behavior among these apes. Another brick fell in the early 1970s, with the demonstration of mirror-recognition in chimpanzees, cleverly documented by Gordon Gallup, a psychologist at the State University of New York, Albany. Many people had suspected that chimpanzees recognized themselves in mirrors, but Gallup offered conclusive proof. He applied a red dot of paint to the heads of a group of chimps while they slept. Upon waking, they were unaware of the paint, that is, until they happened to look into a mirror. At once they stopped, then turned and touched the red splotch on their foreheads to find out what it was.

With the man/animal boundary so deep a part of the Western psyche, it is little wonder that many resist its dismantling on both a logical and emotional level, and with great confusion manifest between the two. Man's ability to exploit the planet, to take of its resources as he needs, and to usurp entire forests and all living creatures therein, rests upon the unwritten assumption that the chasm between himself and all other creatures is impassable. All of modern man's activities operate from the premise that the planet is his to allot into countries, states, counties, and individual plots, because he, unlike other creatures, has been given the twin gifts of reason and expression. By assuming that other animals lack these gifts entirely, man obviates any need to listen to the wishes of the creatures with which he shares the planet. He can therefore proceed comfortably by his own lights, blind to information that is perceived as nonexistent.

To make this progression comfortable, man has wrapped his refusal to recognize the mental continuity between himself and other animals in an elaborate, ostensibly objective, scientific argument. Language, as an instrument of rational symbolic thought, has been the linchpin of that argument. By the early 1990s the boundary between man and animal was still being policed, though a few apes were fixedly gazing into the park with very deliberate expressions on their clearly humanlike faces.

—

Most modern students of animal behavior, whether they work in the field or in the laboratory, count themselves as "empiricists," meaning they accept little as meaningful data that cannot be experimentally reproduced at will. It is because of this empiricist perspective that the study of "human behavior" and that of "animal behavior" have historically been two separate fields, with few overlapping concerns or paradigms. Only a few comparative psychologists have dared to declare that a great deal can be learned about man by studying animals. It is now politically correct to view animal behavior as interesting, even important, but still irrevocably distinct from human action. This distinction is based on the empiricist view that human beings have "minds" and although we cannot know these minds nor even measure them, the minds of human beings can be revealed, one to the other, through language. Because animals lack language as we know it, students of animal behavior conclude that animals therefore cannot tell us "what is on their minds." The empiricist doctrine relies only upon that which can be readily generated and measured at will. It is therefore forced to conclude that since animals cannot manifest minds through language, it is more parsimonious to study their behavior as though it were not generated by mind, intention, or will. However, because humans can—via language—voice expressions of mind, intention, and will, many empiricists accept these concepts as valid constructs for the study of human motivation and action.

The empiricist takes the model of physics as the starting point for the study of behavior. This model suggests that behavior can be reduced to "units" similar to protons, electrons, and neutrons and postulates that only by reducing behavior to its elemental units will we be able to understand its structure. However, unlike physics and chemistry, there is no agreement on what these units should be, nor, indeed, how we should go about looking for them. Nonetheless, the field of animal behavior has long been dominated by a premise called Morgan's Canon. This canon states that animal behaviorists should always seek to explain behavior in terms of the simplest possible processes. Thus, if a dog appears to be hungry, it is thought to be appropriate to define that hunger in terms of percent of normal body weight (that is, a dog that is 80 percent normal body weight is said to be hungrier than one that is at 100 percent normal body weight) or in terms of time since the last food intake. By looking to factors such as "time since last food intake" or "normal body weight," we can avoid attributing eating behavior to "mental states" such as hunger. Of course, as humans who have experienced the state of hunger, or having a desire to eat, we are all quite aware that such a state can, at times, have little to do with either our body weight or the amount of time since we have last eaten. Students of human behavior often view such "irrational urges to eat" as conditioned "bad habits." However, few would deny that a state of hunger in human beings is indeed a mental state and that it can be alleviated by food, as well as by many other activities.

By 1976, the discrepancy between the ways in which animal and human behavior were being investigated became so great that the well-known biologist Donald Griffin began to plead for acceptance of what he called a "common-sense view of animal mind." In a series of three books, published in 1976, 1984, and 1992, he collected reports of animal behavior, not unlike those amassed before him by Darwin and Georges Romanes. Don Griffin echoed their argument that such observations implied awareness and cognition. Since that time, a few researchers have acknowledged that science may well be underestimating the capabilities of animals and have undertaken new techniques for the study of what has come to be called "animal consciousness" and "animal mind." For example, Christophe Boesch, who is studying wild chimpanzees in the Taï forest, has begun to ask questions about the intentional teaching of offspring, particularly as regards the complex skills utilized by this group of apes in nut cracking with hammers and anvils. Robert Seyfarth and Dorothy Cheney have begun to ask questions about whether or not vervet monkeys are explicitly and intentionally attempting to tell each other specific things with their calls, and if so, whether the monkeys themselves are actually aware of what they are doing. Gordon Gallup, a pioneer in the field, whose work predates that of Griffin's, has attempted to determine whether or not various animal species have a concept of self. And of course the language-training efforts of David Premack, Beatrice and Allen Gardner, Duane Rumbaugh (with chimpanzees), Penny Patterson (with gorillas), Lynn Miles (with an orangutan), Lou Herman (with dolphins), Ron Schusterman (with sea lions), and Irene Pepperberg (with a gray parrot) have all suggested that animals must have competencies far greater than currently acknowledged. None of these efforts has yet had a major impact on the way we study human or animal behavior, nor have these two disciplines yet become linked within the field of psychology in the seamless manner that characterizes biology.

———

In a book written in 1863 entitled *Evidence as to Man's Place in Nature,* Thomas Henry Huxley recognized the close anatomical relationship between the African apes and man. The orangutan, an Asian ape, was said to be more distantly related. Huxley concluded that the African apes and man should be placed in the same taxonomic family. Although Huxley's recommendations were ignored at the time, a century later, Morris Goodman of Wayne State University confirmed them at the molecular level. Goodman, like Huxley before him, proposed a reclassification of man and ape at a Wenner-Gren conference in 1962. Goodman, again like Huxley before him, was roundly rebuked for suggesting that the biological data indicate that man and the African ape should be placed in the same family, which would make them as closely related as, for example, the spinner dolphin and the bottle-nosed dolphin.

In the three decades since Goodman's pioneering work, a vast flood of molecular biological evidence has confirmed his—and Huxley's—conclusions. Pressure is consequently mounting to reclassify humans and great apes in a way that reflects evolutionary and biological reality. Humans and the African apes differ in about 1 percent of their DNA. This means that apes are far more closely related to us than they are to monkeys. Yet many people still have difficulty even telling the difference between an ape and a monkey. Monkeys have tails, apes do not. Apes are built very much like us; monkeys have a wide variety of anatomical types, but in general their shoulders, hips, and torso all articulate in a manner similar to other quadrupedal mammals and very different from apes. Apes, like ourselves, are built on the "upright plan," which means that we, and they, tend to orient toward each other in an upright position much of the time. The upright position may be achieved by being suspended from a branch, by sitting, or by standing.

There are differences between ourselves and apes, of course. The most striking is that the size of our brain is three times that of an ape. And then there are the behavioral differences. None championed these more strongly than Huxley himself, even though he argued that man and ape belonged in the same taxonomic family. "No one is more strongly convinced than I am of the vastness of the gulf between . . . man and the brutes . . . for he alone possesses the marvelous endowment of intelligible and rational speech [and] . . . stands raised upon it as on a mountain top, far above the level of his humble fellows, and transfigured from his grosser nature by reflecting here and there, a ray from the infinite source of truth."[13] Huxley's views are widely echoed today. Simply put, they state that language makes it possible for we humans to transcend our biology and thus enter a state of being so different from that experienced by any animals as to render a complete discontinuity between their life experiences and our own.

How did humans come to possess so remarkable a means of communication? Is language a highly developed extension of the communication systems already in use by other primates, or does it represent a complete break, a way of communicating that is so different from that of other species that it makes a completely new form of thought—rational thought—possible? The rationale of ape-language research, of course, is that the cognitive foundations of modern human speech are likely to be found in the great apes, our closest evolutionary relatives, and perhaps in other higher mammals as well. Much of modern linguistics is dominated by the opposite view.

Since the late 1950s, Noam Chomsky, of the Massachusetts Institute of Technology, has argued forcefully for the complete and singular independence of language from all other forms of communication and thought. According to this view, it is the syntactical structure underlying language that, in effect, makes language possible. Without this structure, Chomsky argues, we would be limited simply to expressing linear unrelated concepts and to reacting to single events, one at a time. According to Chomsky, this underlying structure, once it is fully understood, will be found to be identical across all languages and all cultures; that is, it will manifest the universal properties of human intelligence. The ability to relate world events in a causal-linear manner is thought to derive from this structure, and consequently to be as innate a portion of our biology as are our eyes, our hands, or our internal organs. According to Chomsky, "the child approaches the task of acquiring a language with a rich conceptual framework already in place and also a rich system of assumptions about sound structure and the structure of more complex utterances." In Chomsky's view these assumptions "constitute one part of the human biological endowment, to be awakened by experience and to be sharpened and enriched in the course of the child's interactions with the human and material world."[14]

Chomsky argues that one should study language "exactly the same way you'd study an organ, say, the heart."[15] To Chomsky, this means that we need to open up the "language module" and understand its structure, just as we have done with the heart. Just as we have learned that blood flows from the heart to the lungs and back again to be pumped to the rest of the body, we need to understand how syntactical rules regulate the flow of our speech.

When the 1990 Wenner-Gren conference began, therefore, it was with the boundary between humans and nonhumans still being policed and maintained by the scientific community at large, with Chomsky as their guardian.

———

Kathleen Gibson set the stage for the conference, briefly sketching the intellectual questions we were to address. She reminded us that while paleoanthropology had progressed in understanding the evolution of the physical form of *Homo sapiens,* the origin of the mind was still swathed in mystery. Perhaps this should not be surprising, as neuroscientists also have found it difficult to answer a basic question about how the brain functions: Are cognitive skills such as reasoning, language, music, and art controlled by discrete modules within the brain, or by the interrelated agents of a distributed system? Evidence has been adduced for both positions, but nothing is conclusively settled.

Until the structure-function relationship of the modern human brain can be described accurately, any understanding of the dynamics of the evolution of the mind must rely on indirect evidence, such as behavior (that of modern humans compared with African apes, and that preserved in the archeological record). Psychobiologists, such as myself, don't often sit at the same conference table as archeologists, but that lofty aim was why we were in Cascais. We each had a view of the value of our own contribution, as I was soon to learn. To me, it seemed obvious that studies of chimpanzees, humans' closest evolutionary relative, offered us our best insights into the emergence of human mind.

I was therefore incredulous when, on the very first morning of the conference, Iain Davidson, an archeologist from the University of New England, Australia, began to speak, saying, in essence, "Humans are different from apes, and all the 'chimpology' in the world won't tell us anything of interest; chimps [aren't] representational thinkers, therefore we [can't] learn anything useful about language by looking at apes; training chimps to do tricks might be interesting in itself, but it is irrelevant to events in the past; and if you want to learn anything about language origins, the only place to look is at the archeological record; psychologists behave as if prehistoric evidence doesn't matter; finally, the archeological evidence suggests that human language arose very recently, within the last 100,000 years—so much for chimpology."

Iain should have been aware that a word such as "chimpology" would grate on those of us at the conference who had devoted our lives to the study of apes. His assertion that nothing of interest could be learned from chimps about the nature of human language made me feel that an immediate response to such a totally homocentric position was sorely needed. I therefore vigorously objected to the idea that Kanzi had been *trained* to do anything, observing that Iain obviously hadn't had much opportunity to learn what chimps were like. I noted that they easily handled symbolic representation, but I had to stop short of explaining the data that permitted these conclusions. My response had to be brief as Gibson and Ingold were careful to let others have an opportunity to respond, too.

I sat back thinking, "Oh no, another Chomskian! Another scientist whose mind is closed to what chimps can tell us about the human mind." As others spoke I was amazed at the anthropocentric views of many around me. The reaction to language in apes was even more negative than I had feared. For some, Kanzi and other speaking apes were seen as little more than the ghoulish constructions of publicity-hungry scientists who sought to distort the natural order of things by creating apes who could talk. The idea that an ape might desire to learn to talk, for the purpose of communicating his or her thoughts to another species, was not one that many seemed interested in or able to grasp. The idea that we humans might learn something about the origins of our language and minds by observing how apes can go about developing these capacities seemed too remote to hope to present.

I soon learned, however, that Iain was as concerned about people failing to understand his perspective on archeology as I was about having my chimp work misunderstood. Consequently he had sought the very first opportunity to drive home the point that an event as salient as the origin of language, which took place in prehistory, is likely to have left its mark on prehistory, and that this line of evidence had not received sufficient attention from people interested in language origins—particularly psychologists. I learned this only because that evening after supper I decided to muster my courage and go up and speak to Iain directly.

I started out by asking him how he could be so peremptorily derogatory about apes when he knew next to nothing about them. How was it he could so blithely assume that symbol capacities did not exist in creatures before they began to leave permanent statues, paintings, or other artifacts in the archeological record? Much to my surprise, Iain was not nearly as bombastic when approached one on one. I realized that he probably had been so stiff and argumentative because he was in front of the group. It seemed to me that human males often responded that way during initial encounters—just as do chimpanzee males. Unlike every other Chomskian I had met, Iain readily allowed that there was a great deal he did not know about apes, though he held fast to his notion that only man could symbolize.

Although our original intellectual positions seemed very far apart—with Iain proposing a late, sudden origin of language while I favored an early, more gradual efflorescence—we began that day one of the most productive, free-flowing exchanges of ideas I've been privileged to enjoy with a colleague. We both listened, and as a result began to modify our ideas. Iain wasn't afraid to argue forcefully for his views, and neither was I, but somehow this did not keep us from talking to one another, as is so often the case with scientists who disagree. Why not? I think because we each accepted the other's directness, even bluntness, without offense. We did so out of a mutual recognition of the integrity of the other person's efforts. It was a remarkable week, one that may become a milestone in our important intellectual journey toward understanding the human mind—with the human/nonhuman boundary suffering badly.

Fortunately, it was not up to me alone to convince the gathering's skeptics of the range and inventiveness that characterize ape intelligence. I was aware, of course, of chimpanzees'

penchant for using tools, but it wasn't until Bill McGrew item-
ized the different types of tools they use, and in the many different circumstances, that its import became fully clear to me,
and to others, too. "Chimpanzees are the only non-human
species in nature to use different tools to solve different problems," said McGrew, a primatologist at the University of
Stirling, Scotland. "They go beyond using the same tool to
solve different problems (for example, a sponge of leaves to
swab out a fruit-husk or a cranial cavity) or different tools to
solve the same problem (for example, probes of bark *or* grass *or*
vine to fish for termites). Thus, they have a *tool-kit.*"[16]

It was Christophe Boesch, of the University of Basel,
Switzerland, who grabbed the most attention with his reports of
chimpanzees in the tropical rain forest of the Taï National Park, in
the Ivory Coast, West Africa. The chimps of the region exploit a
rich food resource, that of Coula and Panda nuts, which have to
be cracked open to give access to the kernels. The animals gather
the nuts and then place them one by one in a depression in an
exposed tree root or branch on the forest floor (the anvil). They
then pound the nuts one by one with a stone or short branch
(the hammer). A skilled practitioner can garner as many as 3800
calories a day in the nut season.

It takes time to become skilled at nut cracking, however—
often as much as eleven years. The animals' persistence in developing the skill is an indication of the value of the nuts. One of the
great advantages of nut cracking is that it allows a mother to pro-
vision for her offspring. Coula and Panda nuts are common in
Western and Central Africa, but it is only the Taï forest chimp
population that has learned to exploit the resource. This is a striking example of a cultural difference between populations. Even
more dramatic, however, was Boesch's observation of mothers
actively teaching their offspring the skills of nut cracking.

Boesch told us how, on one occasion, he saw Salomé and
her son Satre cracking Panda nuts. Salomé cracked most of
them, and when Satre tried with a partially opened nut, he
placed the nut improperly on the anvil. Before he could strike it
with the stone hammer, "Salomé took the piece of nut in her

hand, cleaned the anvil, and replaced the piece carefully in the correct position," explained Boesch. "Then, with Salomé observing him, he successfully opened it and ate the second kernel."

On another occasion Boesch saw Nina, daughter of Ricci, having difficulty opening nuts using an irregular hammer. Nina kept shifting her position, turning the hammer around in her hand, but still was frustrated by her lack of success. After eight minutes, Ricci joined Nina, who immediately gave the hammer to her mother. Ricci took the hammer and, with deliberate slowness, turned the hammer to the most effective position, cracked some nuts (which she shared with Nina), and handed the hammer back. Nina took the hammer, held it in the position demonstrated by her mother, and proceeded to open four nuts. This example is particularly interesting, said Boesch, because "the mother, seeing the difficulties of her daughter, *corrected an error* in her daughter's behavior in a very conspicuous way and then proceeded to demonstrate to her how it works with the proper grip."[17]

These instances are the first field examples of active teaching—another behavior supposedly unique to humans—in a nonhuman primate population. As Kathleen Gibson kept remarking throughout the conference, "Every time I learned something unique about humans, it wasn't unique!" The gap was closing, or at least bridgeable.

As the week wore on, I grew anxious for an opportunity to show the video of Kanzi: Sydel Silverman finally relented under my repeated requests and promised that I could show my tape. I had lugged an American machine with me, along with all the required adapters, so I could run the tape (which has different specifications from European videotapes). I had been testing the system in my hotel room a few days before I was due to show the tape. Disaster struck. I had left the machine plugged in when a series of power outages—common in the region, I learned later—destroyed a key component in the set. Philip Lieberman, an expert on the structure of the vocal tract and a professor at Brown University, helped me search for replacement parts in the local town. He stripped the set down, put in the new parts, reassembled it all—but it still failed to work. In desperation, I called LRC (the Language Research Center, my lab in Georgia) and asked if a tape could be transformed to European specifications and sent express to Cascais. It arrived late, not until the very last day.

With the tape in hand, I was now ready for Kanzi to demonstrate to a critical but finally open-minded audience just why the remaining bricks in the constantly patrolled man/animal wall should be cast aside.

— 2 —

The Meaning of Words

I FIRST saw chimpanzees in the St. Louis "monkey show" when I was about eight years old. They rode a motorcycle and walked on stilts. They wore clothes and sometimes had temper tantrums in which they began screaming at each other in the middle of the show. I remember thinking that the things they were doing looked like fun, but that the chimpanzees themselves did not look as if they were having a good time, except for the one who was permitted to ride the motorcycle. He went round and round with his chin tucked down and a rather fiendish grin on his face that I can still recall. He jumped the bike through a hoop of fire for the finale—and all this was before Evel Knievel.

I wondered how the chimps felt about what they were doing. I tried hard to broach that question with my parents, but they seemed to think it was an odd question and dismissed it, saying something about how it was not really possible to tell exactly how animals felt about things, since they could not talk.

I also saw the lion and tiger show, in which a trainer with a whip kept ten different large cats on pedestals while each took a turn doing tricks. The most impressive trick was when one of them jumped through a flaming hoop. In contrast to the chimpanzee on the motorcycle, the cat seemed afraid. He readily jumped through the hoop without the fire, but when it was lit, he hesitated and had to be persuaded with a great deal of whip cracking by a trainer who was sweating profusely. At the time, I had no sense of how dangerous apes can be; neither, I think, did the rest of the audience.

The big cats, however, looked perfectly capable of swallowing a man, and they all appeared quite hungry. Consequently, the power the trainer could wield over them with only a whip amazed me. I noted that the trainer of the chimpanzees did not use a whip and the chimpanzees did not threaten him, as did the lions. I never really wondered how the big cats felt about the show, as I did with the chimpanzees; I did wonder why they tolerated being in a show.

My next experience with live chimpanzees occurred when I became a graduate teaching assistant in psychology at the University of Oklahoma, in Norman. Roger Fouts, who had recently transferred to Oklahoma from the University of Nevada, arrived in class one day carrying Booee, who was a little more than three years old. Roger set Booee down on a table and produced a series of objects, including a hat, a shoe, and a ring of keys. He held up each object and asked Booee to name it. The class, myself included, was quite taken aback as Booee formed his hands into signs for the objects, using the American Sign Language (ASL) system. Although he knew the names of all the objects, Booee was really interested only in the keys. He wanted to play with them and kept signing "keys" over and over while holding his hand out for them. While I wondered whether he really knew the other words that he was signing, it seemed pretty clear that he understood "keys."

The time was 1970, and prior to seeing Booee in class I had heard little of the so-called ape-language studies that had begun in the late 1960s. My image of apes—based on the shows I had seen at the zoo and on television—was that of acrobatic monkeys. I thought of apes as smart dogs with hands and goofy faces. But the ape before me was different. He wasn't cavorting on a stage or walking around in a coat with a large hat down over his eyes, mimicking a human character. He was sitting calmly on a table in front of nearly one hundred people and making a different sign each time Roger held up a different object. Roger had shown us the signs in advance, so we could see for ourselves that Booee was correct almost every time.

After watching Booee do this, and noting the expressions on his face as he looked out over the audience, I knew at once that this creature had to possess sentience that all my life I had assumed was to be found only among humankind.

As impressive as Booee's apparent language abilities were, even more intriguing was the nature of the social contract between Booee and Roger. Being the oldest in a family of seven siblings, I had been responsible for much of the caretaking of my younger siblings. This responsibility necessarily gave me a clear understanding of the typical behavior of young children, and the means they use to communicate even before they can speak. Seeing Booee with Roger that day was something like having a professor reveal to you that one of his children was learning impaired with severe speech difficulties and a few minor physical deformities. The real news, however, was not that this "child" was impaired, but that this child was not a human being by physical standards, regardless of whether or not his behavior suggested otherwise. Watching Booee and Roger reminded me of the many times I had taken my younger brothers and sisters on outings as they were learning to talk; Booee was just as playful and inventive as they had been.

I was attracted to psychology, in part, because of the experiences I encountered helping to raise so many siblings. The development of human behavior intrigued me, particularly the effect of rearing on the development of the human mind and intellectual capacities. When I saw Booee I decided to take Roger up on his request for volunteers to work at the "Chimp Farm." This chimp farm, owned by Dr. Bill Lemmon, a clinical psychologist, was located about fifteen minutes outside of Norman, Oklahoma. There approximately thirty chimpanzees resided, along with several species of monkeys, lesser apes, peacocks, pigs, and various other animals.

On my first visit to the chimp farm Roger put me in a very small outdoor cage (two-by-four-by-six feet) with Booee and a bowl full of raisins. He showed me how to get Booee to practice the signs he was being taught. I was asked to hold up the object, then take Booee's hands and place them in the appropriate configuration for the sign that corresponded to the object I had displayed, and then quickly give Booee a reward. There were a number of different objects Booee was learning and my job was to show each one to Booee, in a random order, and then guide him into making the correct sign. Once I could easily guide his hands into the proper place, I was asked to "fade out" the guidance slowly, expecting Booee to do more and more of the work, until finally he would make the sign all on his own whenever I held up the object.

Booee was not very interested in making the signs I was supposed to teach him, nor did the fading technique seem to be as effective as Roger had promised. Playing was a far more effective way to spend one's time in Booee's view, so I soon began playing with him, after each batch of correct signs, to keep his interest. One time he hung by his hands from the top of the cage and did a 360° turn while we were playing. I laughed and, wanting to see that again, held up my hand and spun my index finger around in a 360° arc and pointed to the top of the cage. Booee at once grasped my intent and proceeded to repeat his flip for my benefit. This gesture was not one of Booee's signs. In fact, no one had ever made that sort of gesture or request to him before. He had been asked to sign, but never to do things that entailed his entire body. Yet he was immediately able to comprehend the meaning of my gesture—repeat that flip you did up there.

I felt so taken aback by the alacrity with which he understood what I meant that I simply stood there, rooted in place, staring at him transfixed. The implications of his ability to comprehend such a novel gesture began to race through my mind, leaving me somewhat shaken. Finally Booee decided to take advantage of my rapt state and began to steal the raisins. That brought me back to the necessity of dealing with an ape in the immediate time frame that they, like children, typically insist upon. That night, as I lay awake thinking over all of the things I had seen apes do that day, and how different the "signing chimps" appeared to be from those who had no signs, I knew deep within me that I would work with apes the rest of my life in some manner. For a long time I had felt, in a vague but ever-present way, that there was something I should be about, but that I had not yet discovered it. That night I knew, without doubt, that I had finally found it, and that it was apes.

Booee lived on an island with four other young chimpanzees, Thelma, Cindy, Bruno, and Washoe. When they were not engaged in signing lessons, these four chimps were taken, by boat, back to the island where they played and slept. Roger was attempting to determine if chimpanzees other than Washoe could learn sign language. Since it took a long time, and a great deal of practice for these chimpanzees to learn signs, Roger needed a lot of help. That was why he had brought Booee to class, to help solicit volunteers.

Roger explained to the class that in Nevada he had worked with Allen and Beatrice Gardner, who had initiated an ape-language project in 1966 with a wild-born female chimp called Washoe. By 1972 Washoe had become something of a simian celebrity, with an American Sign Language vocabulary of some 150 signs and an apparent ability to make up novel sentences, albeit short ones. Roger brought Washoe to Oklahoma with him in 1970 and began to teach ASL to three other young chimps, including Booee.

As I began to read the relevant literature, with guidance from Roger, I became even more keenly aware of the implications of human-ape communication. It seemed to offer an insight into the nature of language and even to the essence of humanity itself. There is a long history of this kind of fascination—and specifically in the notion of teaching language to apes. For instance, an entry in the diary of Samuel Pepys, dated 24 August 1661, records his reactions on seeing an ape in London, one of the first to have been brought back to the Western world. He describes the ape (probably a chimpanzee) as being "a great baboone" but notes how humanlike it was. He wrote: "I do believe it already understands much english; and I am of the mind it might be taught to speak or make signs."

Such suggestions were subsequently to be made many times, often as a passing comment of this kind. The most influential proposal, however, was by the pioneering primatologist Robert Yerkes. In a book called *Almost Human*, published in 1925, he noted apes' great intelligence, speculated on their cognitive potential to form language, and recognized the anatomy of their vocal tract as a barrier to speech. "I am inclined to conclude from the various evidences that the great apes have plenty to talk about, but no gift for the use of sounds to represent individual, as contrasted with racial, feelings or ideas," he wrote. "Perhaps they can be taught to use their fingers, somewhat as does the deaf and dumb person, and thus helped to acquire a simple, nonvocal, 'sign language.'"[1]

Four decades later, the Gardners put Yerkes' suggestion to the test. Washoe lived in a trailer in the Gardners' backyard and experienced much of the life of a child, surrounded by playthings and language. The Gardners and their helpers taught Washoe a limited ASL by molding her hands into the appropriate sign in the presence of an object. She was often rewarded for success with some tidbit of food. It was a laborious business, requiring repeated presentation of the object and repeated molding of the hands. A single sign was taught at a time. Slowly she built a vocabulary, rising from four signs after seven months, to thirty after almost two years, and peaking at about 150 after four years. By that time, 1970, Washoe had become too large and boisterous to be easily handled in the Gardners' trailer, so she was moved to the Institute for Primate Studies, in Oklahoma, where Roger continued working with her in the company of the other apes he was hoping would also learn to sign.

At least part of Yerkes' instinct had been correct: An ape could be taught to make and use signs in a simple context. Typically, that context involved requesting some item of food or drink or initiating play—chimpanzees love to chase and tickle. The intuition that chimpanzees "have plenty to talk about" was, however, less substantiated by the Gardners' experience with Washoe. True, food and play are important in chimps' lives, particularly as youngsters. Using ASL in this context, therefore, may accurately reflect what is on the animal's mind. But Washoe's use of symbols did not give insight into what it is like to be a chimp, which was what many people longed for.

Nevertheless, Washoe's acquisition of so extensive a vocabulary of signs was a great achievement, one that seemed to indicate that a chimp could break the human language barrier. For linguists, however, the quintessence of language has become syntax, the underlying structure that orders the utterance of words and imposes overall meaning. The singular focus on syntax is the result of Noam Chomsky's dominance in the field, and his position is simple: Without syntax, language, as we know it, cannot exist. In order genuinely to break the language barrier, chimpanzees therefore would have to demonstrate syntactical competence.

———

During the early 1970s several more ape-language projects were established, inspired by the apparent success with Washoe. At the University of California, Santa Barbara, David Premack taught a female chimpanzee, Sarah, to use plastic shapes as "words." Sarah did not communicate with these shapes, but rather used them to answer specific questions. The questions aimed to determine whether she possessed the cognitive "functional prerequisites" of language competence. For example, Premack tried to determine if she could respond to such typical linguistic structures as negation, class concept, and characterizations of change of state.

Duane Rumbaugh, of Georgia State University in Atlanta, began the LANA (LANguage Analogue) Project at the Yerkes Primate Center. The project, which was closely tied to developing methods for teaching language to severely mentally retarded children, invented a computerized keyboard display of arbitrary signs, known as lexigrams.

Explicitly or implicitly, each of these projects accepted the challenge to demonstrate syntactical competence as a criterion of language competence. Although the details of the results of these projects differed in many ways, some of which were mutually contradictory, there developed a strong conviction during the early 1970s that not only could apes learn symbols, but they also could use them in innovative and structured ways. For instance, the Gardners said of Washoe that she "learned a natural human language and her early utterances were highly similar to, perhaps indistinguishable from the early utterances of human children."[2]

Most of the utterances by the ASL-using apes involved single symbols, but there was a sufficient number of plausible two- and three-word combinations to encourage a sense of rudimentary language. And among these multisymbol utterances there was perceived to be a sufficient degree of structure—with appropriate order for verbs, nouns, and qualifiers, for instance—to encourage a sense of humanlike syntax. It was against this background of rising optimism that I entered Roger Fouts's ape-language project at Oklahoma. After about six months Roger suggested I work with an adolescent female chimp called Lucy. Lucy, who was older than the chimps I'd been working with, lived with clinical psychologist Morris Temerlin and his wife, Jane, in their home not far from the institute. She was being raised in a human home, from birth, as a subject in Bill Lemmon's cross-species rearing experiments. The purpose of these studies was to "determine whether or not maternal behavior was innate." Lemmon, a clinical psychologist and Temerlin's mentor, had become intrigued by the differences between the sexes. In his view, females were biologically programmed to fulfill deep-seated maternal urges. He was therefore interested in determining whether or not these urges were innate in man's closest living relatives.

Lemmon approached this question by placing infant female apes in the homes of his patients and former students. He then instructed the patients to rear the ape as if it were their own child. When the apes reached adulthood, he intended to artificially inseminate them and observe how they cared for their infants; he wanted to see how these apes behaved as mothers, having been reared themselves by a "mother" who loved them, but did not behave as an ape. During the rearing of these female apes, he also used the apes' behavior as a means of "evaluating" his patients, whom he saw on a regular basis.

Lucy was but the first of a number of such "human-reared" apes I was to meet at the Oklahoma Primate Institute. In order to obtain chimpanzees for his studies, Lemmon had developed his own breeding colony of apes. Thus there were many chimpanzees living in social groups who had minimal contact with human beings, apart from the essentials of being fed and having their quarters cleaned. Infants who were born in and remained in these groups were quickly pointed out as being "less intelligent" and "retarded in their development," as contrasted with those reared in human homes. At the time, I did not realize that nearly all my energies for the next years would be poured into trying to understand the differences in behavior and development of these group-reared infants and their human-reared counterparts. The differences were remarkable, but "intelligence," I was soon to learn, is probably the most elusive and detrimental concept prevalent in the conceptual toolkit of modern psychology.

When I first met Lucy she seemed to be more attentive to social interactions with humans and to give crisper signs than the apes in the institute's colony. When Roger introduced me, Lucy retrieved a plastic flower from her box of playthings and offered it to me, just as human children of one or two years do. Object offering passes through a preliminary stage in human children in which the object is first offered and then withdrawn before it is actually taken by the other party. A few months later, the child clearly seeks to transfer objects from his or her possession to that of another.

When I reached out to accept the plastic flower from Lucy, thinking how nice it was of her to offer, she deftly sank her teeth into the back of my hand. Chimpanzee offers, it seemed, were not quite like human offers. I was only later to realize that rather than attempting to give me the flower, Lucy was daring me to take it, albeit in a rather deceptive manner. I would fall for that trick again many times in my ape career, as chimpanzees are of sufficient intelligence to disguise the trick in many ways.

Why should Lucy want to bite me? Obviously I had not tried to bite or hurt her in any way. I just wanted to learn about her. Little did I understand how my presumption in entering her home with no proper greeting or explanation of my mission had been rude from her perspective. Nor did I realize that extending the object without the appropriate facial expressions and vocalizations was a way of determining how well I could judge her intentions. It reminded me of going into other cultures, or even distinct subcultures in the United States, where attempts are made to size up what you know by seeing if you fall for the oldest trick in the book. I flunked the test.

Despite the tenor of our introduction, Lucy and I went on to develop a good relationship. That was because Roger advised me that the best way to make friends with chimps was simply to spend time alone with them. Lucy at the time weighed about seventy pounds and was rather intimidating. I wasn't quite sure what she might do to me when we were alone, as pound for pound, apes are five times as strong as a human male in excellent physical condition. Moreover, their teeth are large and their jaws are able to exert enormous pressure. Since I had already had the end of my finger bitten off by a chimpanzee at the institute, I was a little hesitant to be left alone with Lucy. Yet Roger assured me that this was the best method and that Lucy was far less likely to harm me if I was alone with her than if someone she knew were also there. I did not understand this, as it seemed to go against all reason. Why would a chimpanzee be more likely to hurt a stranger such as myself when someone that it already knew and liked was present? Nonetheless, I accepted the advice and agreed to take Lucy out of her cage, bring her into the main portion of the house, and begin to teach her signs.

It was not as difficult as I had expected. Soon I was even taking Lucy for rides in my two-seater MG all over Norman, Oklahoma. Lucy would point in a certain direction and we would go that way. Sometimes, if I refused to go to a place she really wanted to see, she would take the steering wheel away from me and turn the corner herself. Of course, this could be dangerous, so every moment Lucy was in the car I was always fully prepared to stop immediately in case she decided to drive. She was not permitted to get out of the car, except when we drove to a sixty-acre plot of land outside of Norman, which was owned by the Temerlins. There Lucy could bound out and run free.

I continued to work with Lucy for two years, while also beginning a behavioral project in which I observed four mothers and their infants. I was interested in infant development—the stages of maturation through which youngsters pass. The language project fascinated me, but I recognized that I could not understand what chimpanzees were doing with signs until I first understood chimpanzees far better. What kinds of things did they communicate spontaneously to each other and how did they do so? Were their nonverbal communications simply unconscious emotional expressions as all the literature at the time maintained? If so, why were their signs suddenly "expressions of conscious willed intent?"

It made no sense to me that if you held up an object and taught a creature to place its hands in a certain position when you did so, that somehow this procedure led to the magic of symbolic awareness and the ability deliberately to communicate ideas and thoughts to others. I felt that capacity had to be present already in the creature. Additionally, perhaps because I had a two-year-old son of my own at the time, I was beginning to recognize that something seemed to be missing during my attempts to communicate with the signing apes. I recognized that they could successfully request objects using appropriate signs, but I began to be uneasy about how much they actually comprehended. The unease would always emerge when I tried to engage in true communication, that is, when I asked them a question for which I did not already know the answer.

This missing component could similarly manifest itself whenever I asked the chimps to do simple things, such as to hand me a familiar object. Unsure how to respond, they would often begin to sign back rather than try to do what I asked. Similarly, if I asked them a question, such as "Where shall we go?" or "What shall we do?" they would frequently string together a variety of symbols they knew, particularly ones we had been using recently, in the apparent hope of hitting on one that sounded good to me. It was impossible to avoid the overriding impression that my son was far more deliberate in his attempts to communicate and that his understanding of simple requests and questions went considerably beyond that of the apes. The apes often seemed not to realize that they were being asked a "true" question, that is, an open-ended one that they could answer in any way they wished.

Of course, they were queried all the time with questions

like "What's this?" "Who's that?" "Where's *X*?" but these questions always had a "correct" answer. That is, if asked the name of a person, the chimps were expected to produce the correct name. Questions such as these revealed little about what the chimp itself wanted or thought; they were simply rhetorical questions that required a signed response.

I began to worry about what this meant, and tried to talk with Roger about it on several occasions. Roger was never unwilling to talk about such issues, but he seemed to be of the opinion that it was more or less impossible to "get inside" a chimp's mind and therefore the only reasonable thing to do was to take the sign at face value and focus on questions that had clear answers. Scientifically, I had to agree with his stance, yet I knew that if I regularly took such an approach to my son's language, our communication would soon cease to be a very satisfying affair for either of us. It was important that he understand the things I was trying to tell him, and that he express his thoughts, rather than just answer questions designed to test his knowledge.

Then Pancho came along. Pancho's imminent arrival at the institute caused great excitement because he was said to be a pygmy chimpanzee (*Pan paniscus*), the rare and exotic cousin of the common chimp (*Pan troglodytes*). Pygmy chimps, or bonobos as they are known, are more humanlike than common chimps in many ways, including being more vocal and more communicative and having extremely expressive humanlike faces. They are also less aggressive than common chimps and tend to be very friendly toward human beings, whom they seem to have a remarkable ability to relate to.

Therefore, even though Pancho was an adult male whom no one at the institute knew, I felt confident in taking Pancho for walks around the institute's farm, and I did most of my observations of him outdoors. We even took rides around Norman, as I did with Lucy, and we stopped for root beer, hamburgers, and fries at an A & W. Unlike being with Lucy, I never worried about Pancho grabbing anyone who approached the car, nor about him taking the steering wheel. Often I even took my young son along; Pancho was such a gentleman.

Looking back, it was a crazy thing to do—and probably

illegal—but I developed a very strong sensitivity to Pancho's skills and mine in our social and communicative interaction.

After six to eight months we discovered that Pancho was not a bonobo after all, but a Koola-kamba, which some have suggested may be a naturally occurring chimpanzee-gorilla hybrid, or a chimpanzee-bonobo hybrid. Bonobos and gorillas have many physical similarities (small ears, raised nose, large abdomen, shorter toes, for example). The fact that Pancho had been misidentified for so long is indicative of how few primatologists are familiar with bonobos. Had I realized Pancho's true identity, I'm sure I would have been less willing to be so free with him, thinking he might be aggressive. Fortunately, I had experienced no problems with him. The most significant aspect of our interaction, however, was this: Despite Pancho's lack of signing ability, I could communicate with him as easily and as extensively as I could with Lucy. If learning signing is about language, I mused, and language is about communication, why couldn't I communicate with Lucy more effectively than with Pancho?

When I discussed these issues with Roger—about the apparent lack of comprehension and real communication—he insisted that Washoe understood what was said to her, and the younger apes would develop more and more comprehension as they got older. Okay, I responded, can you demonstrate that Washoe really understands requests that are made of her? Certainly, said Roger, as we talked sitting on the bank across from the chimp's island one day. He turned to Washoe, looked around the island, and noticed that a long rope lay near the center of the island. Washoe, on the shore, was looking up at us. Roger turned to Washoe and signed, "Washoe, go get string there." He gestured in the direction of the string. Washoe looked puzzled, but did begin walking in the direction that Roger had pointed. She looked at a variety of things on the island, touching them and looking back at Roger, as if trying to determine what he meant. She walked right past the string several times and each time Roger signed, "There, there, there (again pointing), there string." Finally, as she again approached the area where the string lay on the ground, Roger began to sign "yes, yes, yes" and nod his head emphatically. As Washoe reached the spot, she picked up the piece of string and was praised fulsomely. "See," said Roger, "she just had trouble finding the string." I was not convinced.

I soon became known at the institute as "the unbeliever." It was a very friendly jest, and we all talked a great deal about my concerns; in fact, I was actually happy to be so labeled. When I compared my two-year-old's language competence with the apes', I still saw discrepancies that made me even more unsure about the strong claims that increasingly were being made for language competence in apes.

First, I didn't have to drill object-sign associations with my son, Shane. Words just popped into his vocabulary. Second, I didn't have to stretch my imagination to understand most of the things he said. They were obvious from the context. I had become uncomfortable and suspicious of some of the rich interpretations of apes' utterances I'd heard or read. No juxtaposition of words was deemed too strange to be interpreted as a reasoned utterance. For instance, the suggestion that the gorilla Koko was making puns and other kinds of word play, and had a concept of death, strained my credulity—even though I wanted to think that an ape was capable of such abstract conceptions. Third, Shane clearly understood more of what was said to him than the apes did. I could ask him to do simple tasks, and he would. When I asked a chimp to do a simple task, even as simple as picking up a specific object in front of her, there was often puzzlement—just as Washoe had experienced with finding the string. It was clear to me—as it is clear to any parent—that Shane's ability to comprehend language developed ahead of his ability to produce language. This seemed not to be the case with chimps.

I began to form the notion that comprehension, not production, was the central cognitive feature of language, particularly language acquisition. Comprehension is much more difficult to quantify than production of words, and so linguists had paid little experimental attention to it. In any case, the hegemony of syntax in linguistics ensured that production was held to be the defining characteristic of language competence. For the most part, ape-language researchers accepted what linguists said, and then strained to satisfy their criteria. Seeing this, I became discouraged at the prospects of moving ape-language research forward, and opted instead to devote the rest of my time at Oklahoma to studying infant development. I wanted to document as extensively as I could, the type of things that apes could communicate by using their accepted nonverbal system of glances, gestures, postures, and vocalizations.

—

I met Duane Rumbaugh, of Georgia State University, at the 1974 meeting of the International Primatological Society, in Kyoto, Japan. By that time, Duane was two years into his ape-language project with the chimpanzee Lana, at the Yerkes Regional Primate Research Center of Emory University in Atlanta. In Kyoto I gave a paper on some work I had done earlier with Lucy, so Duane knew I'd had an interest in ape-language research. He also knew I was skeptical of some of the rich interpretation that was being made of apes' language competence. Nevertheless, he invited me to a symposium he was organizing on ape language, to be held at the Southeastern Psychological Association meeting, which was to take place in Atlanta the following year. The paper I gave in Atlanta compared my experiences with Lucy and Pancho, and stated clearly that I saw no communicative advantage being bestowed on Lucy by her ability to sign. I was therefore clearly stating my position as an "unbeliever."

Six months later Duane called me in Oklahoma with an offer of a postdoctoral position at Georgia State University. This position would permit me to study at the Yerkes Center. I accepted at once, because Yerkes is internationally recognized as being preeminent in primate studies—and it had bonobos. To Duane's disappointment, I refused to work on the Lana project, and concentrated instead on studying bonobo behavior and comparing it with that of common chimps. Gradually, however, I was drawn into the language project, partly because of various problems that had developed when Lana was moved to a new, improved facility; Duane asked me to help sort them out.

Initiated in 1971 with the explicit aim of developing systems with which to teach language competence to severely mentally retarded children, the project had by far the most sophisticated means of symbol manipulation of all ape-language projects. The symbols were arbitrary geometric forms, which were displayed on a computerized keyboard. Lana activated a symbol by touching a key, which then lit up and was projected on a screen. As with all the ape-language projects at this point, Lana's communication was principally about food. Unlike what happened in other projects, however, she was required to build a sentence in order to obtain food or some kind of play activity: for example, *please machine give piece of orange* and *please machine make music.* After four years Lana had an extensive vocabulary with the system, some one hundred symbols, and she had produced a small but significant number of novel sentences. Most of her utterances, however, were of the sort just mentioned. My first impression of Lana was that she knew what she was talking about. I'm sure that the sentence structure she had been taught to produce on the keyboard encouraged that belief. After all, linguists insisted that word order—syntax—was the key to language, and Lana was producing such order. However, I soon formed the same unease with Lana that had surfaced at Oklahoma concerning the chimps' ability to comprehend the words they used. Lana was producing sentences that were comprehensible to me as English construction, but, I wondered, did Lana understand the meaning of the words she was using? When I asked her to do things (such as give me an object), using the same vocabulary of words she employed in her sentences, she could not respond reliably. In an attempt to respond, she would produce inappropriate stock sentences and sometimes nonsentences. She seemed to be searching around for something that would satisfy me, but she wasn't sure what it was. Lana, like Washoe and the other chimps, appeared to be productively competent (using words to request things) but not receptively competent (comprehending what was said to her).

I began to wonder whether there might be several aspects to what we call a word. We assume with human children that the learning of a word includes its comprehension. It seemed to me that productive competence and receptive competence might be discrete cognitive abilities and, in apes at least, might have to be taught separately. Perhaps ape-language researchers had made the mistake of assuming that, as with human infants, once an ape learned a word it also understood its meaning. Perhaps making the leap to search for signs of syntax was not only premature, but also irrelevant to the core of language.

With these kinds of questions still somewhat inchoate in my mind, I began voicing my concerns to Duane. At first he was unconvinced by my suggestions that he and other researchers were making assumptions about the language competence of apes, and were falling victim to a rich interpretation of the apes' utterances. But he listened. We agreed that I would assist him with a new project, with two young male chimpanzees, Sherman and Austin, with a slightly different focus. Beginning in 1976, I established a much closer physical proximity with the apes, interacting with them in a social, preschool-like setting. This would emphasize communicative needs rather than promoting teaching efficiency. The distinction, I believed, was fundamentally important. Further, unlike all previous ape-language projects, this one would not have as its goal the production of word combinations or sentences. I wasn't in search of the linguists' holy grail. I was going to focus on words: What does a word mean to a chimpanzee, and how can we find out?

Just as I was embarking on the Sherman and Austin project, in 1976, a storm was gathering over ape-language research as a whole. The storm clouds rolled in from two directions: From one, linguists poured scorn on the validity of the research and questioned the competence of the researchers. From the other, a prominent ape-language researcher—Herbert Terrace—declared that he and his fellow researchers had been mistaken in believing that apes had acquired language. By the end of the decade ape-language research was completely engulfed in the resulting turbulence, and as a field of study was all but destroyed.

For some time during the mid-1970s, Thomas Sebeok, a linguist at Indiana University, had been expressing strongly negative views on ape-language research. And in May 1980 he organized a conference under the auspices of the New York Academy of Sciences, which made his position brutally clear. The conference was called "The Clever Hans Phenomenon: Communication with Horses, Whales, Apes, and People." Clever Hans was a horse that performed apparently amazing arithmetical feats in nineteenth-century music halls. The putative equine genius would tap its hoof the appropriate number of times in response to a puzzle asked of him by his owner, Wilhelm Von Osten. Unbeknownst to the innocent Von Osten, however, a barely perceptible movement of his head when Hans reached the right number cued the horse to stop tapping. The phenomenon, besides revealing how extremely sensitive animals can be to body language, showed how easy it is for a human to cue an animal unwittingly, thus shaping its behavior. Useful in the music hall, cuing can disrupt the objectivity of research on animal behavior.

Ape-language research is prey to the Clever Hans effect, or worse, opined Sebeok. The results of the research are to be explained as a result of "unconscious bias, self-deception, magic, and circus performance,"[3] he said six months before the New York conference. Conference participants included experts on the tricks of animal performers and a magician, the Amazing Randi. It was "a celebration of deception in all its varieties," a reporter for *Science* noted. "It was amazing that any ape-language researchers should even have considered stepping into such a lions' den."[4] Many such researchers, having accepted a preliminary invitation, dropped out when the tenor of the gathering became obvious. Step into it Duane and I did, however, mainly at his insistence. He wanted to demonstrate that not only did we have courage, but we also had good data. We would stand on our data, our research methods, and our convictions.

It didn't matter how good our data were, however. The atmosphere was so very negative that the issue was effectively prejudged. For instance, the first speaker, Heini Hediger, of the

University of Zurich, declared it "amazing" that anyone would seriously consider that nonhuman animals might display elements of human language. In his presentation, Sebeok suggested that funding for the work should be halted, and perhaps dispersed to more worthy causes, like cancer research. There was even a move, fortunately thwarted, to have the conference vote for a ban on the research. This was reminiscent of a ban on the study of the origin of language, instigated by the Linguistic Society of Paris, in 1866.

Sebeok, with Jean Umiker-Sebeok, had circulated a manuscript that was highly critical of all ape-language research and contained much that was inflammatory. "Thus we find the ape 'language' researchers replete with personalities who believe themselves to be acting according to the most exalted motivations and sophisticated manners, but in reality have involved themselves in the most rudimentary circus-like performances," they wrote, a remark that *Science* noted was "hardly best calculated to brush up their colleagues the right way." *Science* was right. We were further brushed the wrong way with the following: "The principal investigators themselves, of course, require success in order to obtain continued financial support for the project, as well as personal recognition and career advancement. . . ."[5]

At a press conference at the end of the meeting, Sebeok expressed his view most stridently of all: "In my opinion, the alleged language experiments with apes divide into three groups: one, outright fraud; two, self-deception; three, those conducted by [Herbert] Terrace." Pressed by reporters, Sebeok declined to present any evidence to substantiate his accusation of fraud. Whatever motives lay behind such remarks, and behind the initiative for a vote to ban the research, they hardly seemed to be based in science. A "Talk of the Town" column in the 26 May 1980 issue of *The New Yorker* began a commentary on the conference this way: "Can scientists speak to apes? Can apes speak to scientists? Can scientists speak to scientists?" Its answer was that "the jury is still out on all three questions." The answer to the third question, of course, is: Not when the discourse has clearly gone beyond the realms of science.

In my presentation I responded directly to the Sebeoks' lengthy critique, embodied in their circulated manuscript. For instance, they suggested that because ape-language researchers are critical of one another's work, all the work should be dismissed as inconclusive. This struck me then, and still does, as a curious way of looking at science. There are always disagreements among scientists in new, fast-moving fields. And if the conflicts in my field seemed unusual in any way, one can see that this stems from the perhaps inevitable way in which results were sensationalized in the popular press. The Sebeoks asserted that because the apes in the studies work better with some humans than others, this indicates that some people are better at cuing than others. We pointed out that anyone who has ever worked with animals, particularly ones that are sensitive to social interaction, knows that the best work flows from a trusting relationship. To invoke cuing as the sole reason for such differences is to display an ignorance of human/animal interaction.

The Sebeoks suggested that circus trainers, magicians, and others practiced in the art of purposefully manipulating animal behavior to create the illusion of humanlike activities should be allowed to investigate ape-language projects. "Circus performers and gurus are not qualified to evaluate the serious efforts of scientists," we said in our reply. "To turn to psychics and magicians as a cure for unintentional cues by ape researchers is anti-science and anti-intellectual." I would be the last to deny that cuing has happened in some of the ape-language projects. Chimpanzees are extremely smart animals, and can pick up on the slightest traces of approval or disapproval in a human's facial expression or posture. But it is facile at best to dismiss *all* ape-language research as beset by cuing, or to assume that scientists are unable to eliminate cuing possibilities by scientific means.

Perhaps Sebeok's position is best illustrated by his remark to a reporter for the British science magazine *New Scientist*. Asked what kind of evidence would convince him of the validity of some kind of cognitive substrate for language in apes, he replied: "Facts do not convince me. Theories do." Sebeok apparently overlooked the fact that ape language is helping us formulate new theories of language acquisition that have an evolutionary basis.

Besides Duane and myself, the only other ape-language researcher to attend the Clever Hans Conference was Herbert Terrace. Unlike us, however, he was there to concur with Sebeok: The languagelike utterances of apes were a form of Clever Hans effect, he said. "Nim had fooled me," he later wrote.[6]

Project Nim started in late 1973, shortly after Terrace collected the infant chimp named Nim from the Institute for Primate Studies, in Oklahoma. Terrace's initial plan was to demonstrate conclusively what he considered other ape-language researchers had concluded only anecdotally. For instance, the Gardners stated: "The most significant results of Project Washoe were those based on comparisons between Washoe and children, as in the use of order in early sentences."[7] The demonstration of syntax, the holy grail of linguists and, by default, ape-language researchers, was also Terrace's goal: "knowing a human language entails knowing a grammar," he said.[8] Terrace's hope was to demonstrate that Nim knew grammar, by videotaping and analyzing as large a proportion of the chimp's utterances as possible over a period of years, as he learned and used sign language.

More than sixty sign-language teachers worked in Terrace's project in its four years, and Nim learned just as other chimps had. He eventually had a vocabulary of more than 125 signs, and produced many double-word combinations. As the project proceeded, Terrace and his colleagues grew more and more confident that Nim's utterances displayed syntax. For instance, when Nim used *more* in a two-word combination, it appeared in the first position 85 percent of the time, as in *more banana* and *more drink*. Similar appropriate word placement occurred with *give* (as in *give apple*) and with transitive verbs (such as *hug, tickle, give*) when combined with *me* and *Nim*. "The more I analyzed Nim's combinations, the more certain I felt we were on solid ground in concluding that they were grammatical and that they were comparable to the first sentences of a child," said Terrace.[9]

In 1977, after Nim returned to Oklahoma, Terrace and his

colleagues had time to scrutinize the hundreds of hours of videotape they had amassed. The tape had captured twenty thousand of Nim's utterances, half of which were combinations of two words or more. The more Terrace studied these utterances, however, the more his previous certainty eroded. What had seemed like grammatical conversational utterances in the intimacy of the teaching room dissolved into zombielike imitations on the video screen. Terrace had been troubled early on by the lack of increase in the length of Nim's utterances, something that is a natural part of language-learning in children. Nim did produce some long utterances (such as *give orange me give eat orange me eat orange give me eat orange give me you*), but they were usually nonsense, and contained no more information than short utterances. However, the insight that to a great extent Nim was imitating his teachers, rather than spontaneously conversing, was the most telling blow. By the time he was four years old, Nim fully or partially repeated what had just been said to him. And for more than three-quarters of the time, Nim's utterances were preceded by a teacher's utterance.

Terrace came to realize that Nim's use of signs focused on requests for food or play activities. He also realized that Nim frequently threw in "free words," such as *Nim, me, hurry, give,* and *more,* which often added an air of "sentence structure" to an otherwise simple "food" or "tickle" request.

As a result of the analysis, Terrace published a paper in the 23 November 1979 issue of *Science,* titled, "Can an Ape Create a Sentence?" with Laura Petito, Richard Saunders, and Tom Bever as co-authors. The answer to the title's question was a resounding No! "Objective analyses of our data, as well as those obtained from other studies, yielded no evidence of an ape's ability to use a grammar," they concluded.[10] "Sequences of signs, produced by Nim and by other apes, may resemble the first multiword sequences produced by children. But unless alternative explanations of an ape's combination of signs are eliminated, in particular the habit of partially imitating teachers' recent utterances, there is no reason to regard an ape's utterance as a sentence."[11]

Terrace clearly did a thorough piece of work in scrutinizing the reality of Nim's utterances, recognizing that what was inferred as syntax was in fact an illusion of the teaching method and the assumptions of the research. (However, Terrace was incorrect to say that Nim fooled him; in reality, Terrace fooled himself, because the system effectively taught Nim to imitate.)

The combined effect of Sebeok's Clever Hans Conference and Terrace's *Science* paper was, however, to instigate an extremely rapid and violent swing of the pendulum. Ape-language research went from being a field of perceived intellectual excitement and public acclaim to one that, at best, should be viewed askance. Suddenly, it became extremely difficult to have research papers reviewed, let alone published. And funding for most of the major projects virtually dried up. Fortunately, our work had just received a five-year renewal of funding as the storm broke, which allowed us to ride out the storm.

As I watched the credibility of the field crumble, I worried about a complete loss of objectivity about the research. With Duane I wrote a note to *The Psychological Record*, pointing out the danger of overreaction to Terrace's work: "If . . . the general scientific community now concludes (a) that apes are just like other animals, (b) that all animals are very different from men, and (c) that problems only arise when we attempt to study animal cognitive processes, Terrace's personal step forward will become a large step backward in man's struggle to understand the emergence of human uniqueness."[12]

As important as Terrace's work had been, I recognized that it was not fundamental to real progress in ape-language research. It was, of course, important to demonstrate that multiword utterances were often the result of imitation. But this still left the field firmly in the hands of linguists, to whom syntax is sacred. By the time of the Clever Hans Conference and the publication of Terrace's *Science* paper, I knew that this was misguided, particularly as applied to ape-language research. As a result of my earlier concerns over apes' comprehension, and my more recent project with Sherman and Austin, I knew that we had to shift our attention away from sentences and turn it to words.

Because researchers saw that apes built vocabularies so easily by associative learning of individual symbols, there was little thought about what the symbols might mean to the apes. Researchers simply leapt ahead to look for signs of syntax in word combinations. In doing so, they were assuming that when an ape learns a word, it attaches the same linguistic representational aspect to it that children do when they learn a word. The central issue is this: Does the ape *know* that the symbol can stand for an object that may be absent?

There are four components to linguistic representation: (1) an arbitrary symbol that stands for, and can take the place of, a real object, event, person, action, or relationship; (2) stored knowledge regarding the actions, objects, and relationships relating to that symbol; (3) the intentional use of that symbol to convey stored knowledge to another individual who has similar real-world experiences and has related them to the same symbol system; and (4) the appropriate decoding of and response to the symbol of the recipient. A word comes to represent an object, separate from it in place and time.

It was not clear that Washoe or Lana, or any of the other signing apes I had encountered, possessed full linguistic representation, or referential, abilities. They could use a symbol, in what was often a contextually cued situation, to request an object or activity. But they were usually unable to decode the symbol when it was used by a human in a simple request. As a result, the complexity of communication achieved by these apes was no greater than the basic communicative level of chimpanzees who were not language trained. This absence of full comprehension in language-trained apes was, I felt strongly, a more fundamental criticism of ape-language research than the absence of syntax, as demonstrated by Terrace. Cooperative comprehension is fundamental to language, and two-way communication that does not reflect comprehension is not language, no matter what other attributes it may possess.

I had reached these conclusions as the storm over ape-language research was about to break, and assembled them in a paper titled "Do Apes Use Language?" in *American Scientist,* with Sarah Boysen and Duane Rumbaugh as co-authors. Because of extraordinary delay in publication, the paper came out two months after Terrace's *Science* paper, but four months prior to the Clever Hans Conference. We concluded the paper with the following: "Experimenters must stop looking for superficial similarities between apes and children and must instead investigate the cognitive competencies that underlie symbolic processes."[13] That was precisely where the project with Sherman and Austin would take us.

— 3 —

Talking to Each Other

S HERMAN and Austin were two and a half and one and a half years old, respectively, when the Animal Model Project commenced, in June of 1975. The primary goal of the project was to elucidate the processes of language acquisition in apes and compare them with the phenomenon of spontaneous language acquisition in human children. This goal encompassed practical and theoretical issues. First, it continued and extended the effort to develop language-training techniques that might help severely mentally retarded children. Duane had initiated that endeavor at the Yerkes Regional Primate Research Center in 1970, with the Lana project. Second, it asked, *in what sense* can a species other than *Homo sapiens* develop language?

The unease I had experienced over the strong claims for language competence in the chimpanzees I was most familiar with—Washoe, Lucy, and Lana—encouraged me to guide the investigation with Sherman and Austin in a very different direction. The goal of most researchers in the field had been to determine whether apes *have* language, in much the same way as you might determine whether they have a thumb or a stomach. As we saw earlier, it was expected that if apes do have language, its presence would be revealed by the animals' innate syntactical competence, a putatively genetically determined ability to order the symbols in multiword utterances.

My goal was at once to be more modest and more ambitious than seeking signs of syntax: I planned to focus on words, not

59

sentences. More specifically, I was interested in the animals' ability to comprehend and communicate. I had no ambitions to instill in Sherman and Austin an impressively extensive vocabulary. Nor would I spend time encouraging multisymbol utterances. Instead, I was reaching beyond these staples of ape-language research, seeking to touch the essence of language: the ability to tell another individual something he or she did not already know. I wanted Sherman and Austin to use symbols referentially with each other, in true, humanlike communication.

The journey toward that goal turned out to be longer and more arduous than I had expected, and at every step of the way I encountered problems, primarily because of unsubstantiated assumptions I made about what Sherman and Austin would be able to do as we progressed. I came away from the experience— one I felt I had jointly shared with Sherman and Austin—with a better understanding of the nature of language and its acquisition. The results of the project also advanced ape-language research in a fundamentally conceptual manner.

The unfolding of the Animal Model Project happened to parallel in time the rising fomentation over the validity of ape-language research, which I described in Chapter 2. By the time Herbert Terrace published his influential *Science* paper, in November 1979, and the Clever Hans Conference of the New York Academy of Sciences had taken place, in May 1980, Sherman and Austin had achieved a level of language competence—in the sense of true symbolic communication—that far surpassed that of any of the apes so frequently cited by both proponents and opponents of ape-language research. I had been working with them for five years at that point. Ironically, because I had eschewed syntax as a goal in my project, Sherman and Austin received not even as much as a footnote in the debate.

———

From the very beginning, the two chimps were very different from each other. Sherman was, and has remained, the physically bigger of the two. Partly as a result of this, Sherman has the ability to be the dominant individual whenever he chooses. Just as striking, however, are their personality differences. Sarah Boysen once aptly described Sherman as the football player and Austin as the stamp collector. Sherman has always been extremely active and reactive and always in the middle of things when some commotion is going on. When he was young he was constantly hurling himself into my lap and leaping onto my head—I suffered a perpetual sore neck as a result. And when he was older I had to tape thumbtacks onto the back of my hands to discourage his rough, albeit friendly, play-bites. When Sherman gets mad he erupts volcanically, with impressive displays, rushing around, puffing his hair out, banging things, and even slappings or pushing me. But it is all over very quickly. You know where you are with him.

Austin is much quieter, gentler, and slower to react. But when he does get angry he is much more dangerous than Sherman, because he often seems to slip beyond his own control. On occasion he has pulled me off my feet and thrown me to the floor so forcefully that it is difficult to get up afterward. I often feel sorry for Austin, because at two months of age he apparently couldn't digest his mother's milk and was failing to thrive. He was therefore separated from her and as a result displayed a behavior common to young chimps who suffer early separation: Whenever he was distressed he rocked rhythmically from side to side while holding his blanket. Fortunately, as an adult he has outgrown this behavior. I have no idea why, but when Austin sees a picture or a doll-like figure of a human or chimpanzee infant, he tries to destroy it, and becomes agitated if he is thwarted.

In the years I've known the two chimps, I've never seen anything that would indicate a significant difference in their basic intelligences. Nevertheless, it is interesting to see how their personalities are reflected in the way they approach tasks. For instance, Sherman always prefers to communicate with gestures and other nonsymbolic means if he can. Austin is a keyboard individual. And Sherman is much better at tasks that require participation, while Austin excels at tasks requiring close observation and attention. These differences, and differences in

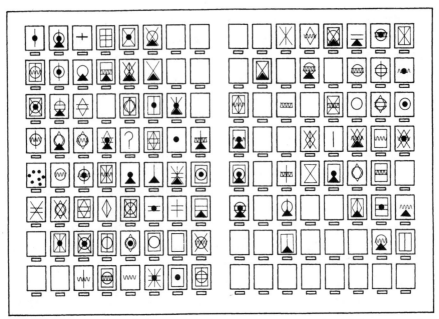

All of the lexigrams used by Sherman and Austin, just as they were arranged on their original keyboard.

the mistakes each makes with symbol-use from time to time, have persuaded me that we shouldn't talk facilely about *the* chimpanzee mind. Surely, Sherman and Austin share something of a basic chimpanzee view of the world, but their individual experience of it is undoubtedly very different.

The communication system we used with Sherman and Austin was a modified version of the computerized keyboard that Duane had developed for the Lana project. Symbols that represent objects or activities are arbitrary geometric forms, based on the Yerkish language that was invented by Ernst von Glaserfeld. Their keyboard eventually grew to 92 symbols.

Early on, we activated only a few symbols on the keyboard; as their vocabulary slowly grew we added more and more. When pressed, the key lit up, and was very obvious. To avoid the problem of the chimps simply learning the position of a key rather than the symbol on it, the symbols were randomly re-assigned new positions after each keyboard use. The chimps had to search carefully for the symbols they wanted to use. Even though Sherman and Austin have never shown any extensive comprehension of human speech, we decided from the beginning to use speech as well as symbols to get them to do things. They were evidently very sensitive to affect in the voice, and responded appropriately to emotions it conveyed.

The beginning of the journey toward true communication started with what everyone had assumed was a simple task—teaching the names of objects by association. For Washoe, Sarah, and Lana, this teaching of "names" had shared the important common procedural elements of holding up an object, an apple for example, and then encouraging and/or helping the ape to make the sign or to select the correct symbol. Once the ape could do this on its own, additional signs or symbols were introduced to build vocabulary. It was assumed that teaching words in this way was a rather simple process, if somewhat time-consuming. It was also assumed that such training basically depended on forming conditioned stimulus response associations between items and symbols. The real test of language, it was said, would come when one looked to see if apes could put together sentences from the words they had learned.

While this sort of word learning may lead to sentence production in adults who are learning a second language, it is not at all the way children go about learning their first language. However, at that time, very little research on how children actually learned words had been done. It seemed reasonable, therefore, to start by teaching apes word-symbol associations. It had also seemed to work.

Unfortunately, the process did not work with Sherman and Austin. With only one symbol available to be selected, Sherman and Austin learned relatively easily, as would be expected. If we held up a banana, they selected the banana symbol, as that was the only one available. With two, a lot more practice was required, but eventually they succeeded. Beyond two, however, the chimps became hesitant and failed to improve, no matter how much practice we gave them. The chimps were clearly puzzled as to how to proceed. I was puzzled, too, because Washoe and Lana had encountered little difficulty in learning associations between objects and symbols or signs.

I watched videotapes of training sessions to see if I could figure out what we were doing wrong. The problem was that I had not stopped to ask why a chimpanzee who had had no previous language training would know that symbols encoded anything. I expected the chimp to make an association between the object and the symbol—and thus know the object's name. Instead, as I learned from close scrutiny of the videotapes, Sherman and Austin were paying attention to the symbol and my subsequent action—that is, whether or not I gave them a reward.

No wonder we were all confused—humans and chimps alike. We all had different views about what was going on. As teachers holding up objects, we assumed that the object we were showing the chimp would serve as the "stimulus object" since it preceded the response. The chimps, however, were assuming that the symbol-key they selected served as the "stimulus" for us to give them food. They looked for a link between the symbol they depressed and whether or not we elected to give them food—and if so, what sort of food. They paid no attention to the "stimulus item" we displayed. Indeed, they paid no attention even when we attempted to cue them by pointing back and forth repeatedly between the stimulus item and the correct lexigram.

In hindsight, this seemed obvious. Why should they care what we were showing them? They cared more about whether we were tickling them, chasing them, feeding them, and so forth. Once it was apparent that the chimps were attentive to the consequences that followed their symbol production, rather than to the stimulus conditions that preceded it, it suddenly dawned on me that the Gardners had essentially taught Washoe "names" by giving her the object or action *after* she made the sign. Thus, if she signed *tickle*, they tickled her, if she signed *banana*, they gave her a banana. Similarly, when Lana depressed *Please machine give piece of banana*, she got a banana; when she said *Tim tickle Lana*, Tim did so—at least he did so at first while Lana was initially learning. However, this fact had not been pointed out in articles. Instead, it had been emphasized that Washoe and Lana had learned the names of things by being shown the objects they were to name.

After it became clear that the contingencies which followed symbol use made all the difference, we switched our teaching approach. Now we showed the chimp a banana and when he selected the banana symbol, he was permitted to eat the banana.

This sort of transaction took the form, "I show you X, you select the lexigram that goes with X, and then I give you X," and has been characterized as a request task. It has the superficial appearance of language, in that an ape who can make many different signs or select many different symbols when shown various objects seems to be able to ask for a wide variety of things it wants, such as foods, tickling, and so forth. What the chimp does, however, should not be confused with knowing names. Even though the ape appears to "name" something that it is being shown, it is really selecting the symbol because it anticipates that it will receive the object. This is true whether the teacher is holding up a banana, preparing to groom the ape, or getting ready to open up a door or box. In each case, the chimp is selecting the symbol or making the sign on the basis of what it anticipates will happen afterward.

The switch in training procedure produced rapid learning. Within a few days Sherman and Austin were correctly requesting a number of different foods. Now symbol-object pairings were coming as easily to them as had been described to be the case for Washoe and Lana. However, if Sherman and Austin were asked to name something when *all* expectancy of being able to receive that item was removed, they began to evidence confusion once again.

This was the first realization on my part that holding an object up and having a chimp produce the right sign or symbol did not mean that the ape knew the name of the object. It only meant that under certain conditions, the chimp knew what to do in order to obtain the object. In contrast, once a child knows words it seems able to do quite a bit more with them, even before it begins to form sentences. Certainly a child does not limit its word use to occasions set by the parent who is holding up items and asking for their names, or waiting to open doors or tickle the child until it says the appropriate word.

From these early experiences I learned three important lessons. First, chimpanzees do not necessarily learn object-symbol associations easily through practice and repetition, as had been supposed. Second, I did not have to worry about cuing Sherman and Austin into giving erroneously positive results; cuing, if it was happening, was clearly ineffective as a means of teaching. Third, the simple relationship that was assumed in the object-symbol association was not simple at all; such associations could be of several types, each with a different implication for inferred language competence.

Also from these experiences flowed a means of communicating wishes, which Sherman and Austin unexpectedly developed on their own. Our technique required either me or another teacher to put some item of food in a dispenser linked to the keyboard. The chimp's task was to hit the key that corresponded to the food, and thus receive it. Very soon the chimps began to pay close attention to the food I was about to select, and would become impatient if they thought I was too slow. Before long they began to hit a key *before* I selected the food. Were they trying to control my behavior, getting me to select foods they liked? It seemed so. They were very attentive, and if they hit a symbol for a food they knew was in the refrigerator, they would stare at the refrigerator after hitting the key, apparently waiting for me to respond appropriately. As they always chose favorite foods, like M&M's or juice, and never water or chow, the behavior did have the appearance of intention about it.

Initially I discouraged this behavior, but then I realized it was indeed communicative and as such was important in our overall goal. We therefore incorporated it into our teaching regime. Once the chimps learned to use specific symbols for specific foods in this way, their progress improved dramatically. We were pleased with the chimps' headway, and added more and more symbols, including some nonfood items, such as "tickle" and "out." The value of communication as a motivator to the chimps' learning was plain, and I often reflect on the benefit of abandoning strict experimental procedure when an opportunity for an unexpected breakthrough offers itself, as it did in this case.

———

The ability to request is just one of three elements that must combine to produce true communication between individuals. The other two are (1) the ability to name objects and (2) a comprehension of symbols as referents of objects. We were unsure of the precise learning path we needed to take to have Sherman and Austin communicate symbolically with each other, but we knew they would need to be able to use their symbols without expecting some particularly beneficial contingency to follow each time. After all, if Sherman was to do something as simple as to ask Austin to give him a banana, Austin would have to understand that "banana" meant the specific fruit, even though he was giving one rather than getting one. Separating "names" of things from the contingencies associated with the learning of those names proved to be more difficult than we had anticipated.

Knowing how to use the symbol "banana" as a way of getting someone to give you a banana is not equivalent to knowing that "banana" represents a banana. (It is not obvious that this distinction had been seriously dealt with by other ape-language researchers.) Full communication would require that the chimp be able to use the symbol "banana" without expecting to receive one. We hoped we might simply reverse the request task to achieve this understanding on the part of Sherman and Austin. We held up a food item and lighted the "?What this" symbols. We didn't expect them to know immediately the intent of what we were saying, but we tried to make it clear by nonverbal means. If the chimps replied correctly, we gave them another food item as a reward.

Initially, both chimps selected the appropriate symbol when we held up specific food items, but they then reached out for the food, expecting to receive it. They were puzzled when they were given something else instead, and continued to gesture for the original food item. Eventually, they became upset at what

was happening, not because they were being deprived of food—
they weren't—but because their expectations were being vio-
lated. Soon they stopped producing the appropriate symbol for
the food item being displayed.

We pondered how we might make the distinction between
requesting things and naming things more obvious to the
chimps, and decided to make the food items we displayed
nonedible. We did this by coating the food with plastic resin
(the sort you see in the windows of restaurants in Japan). We
hoped that by removing the possibility of consuming the food,
we would also remove the chimps' expectation that they would
receive it when they selected its name.

This procedure proved to be futile as well. When Sherman
first saw me hold up the plastic-coated M&M's, he hurried to the
keyboard, said "M&M," and held his hand out to receive the
candy. When I gave him one, he popped it into his mouth, got a
strange look on his face, and spat it out. He then asked for
another one. I gave him another one and he did the same thing.
After a few more trials, he looked at me and said "banana"
instead. When I gave him a piece of plastic-coated banana, he did
not even bother to taste it before throwing it on the floor and
stomping on it. This didn't help because there was no way to
"open" this banana. When it became apparent that none of these
plastic-coated foods was going to be edible, Sherman refused to
use his keyboard to name or request them. Austin deemed the
task equally meaningless and both started pointing at the refrigerator and the door to communicate their wishes to me in a more
direct manner.

Sherman and Austin finally learned to name things inde-
pendently of any event that followed naming by a procedure
known as "fading." It begins like a request regime, with a food
item being held up and given to the chimp after he lights the
correct symbol. Fading refers to the fact that the size of the
food item given to the chimp gradually diminishes, while the
size of the item he is being shown does not change. At the same
time, we lavish great praise for correct answers, and give a size-
able portion of a reward food item. For example, if the food we
wanted them to name was "sweet potato," we gave them a

smaller and smaller piece of sweet potato each time they did so, along with a large piece of some other food, to indicate we were still pleased. This additional food was the same whether the food they were asked to name was a sweet potato, an M&M, or soycake. The procedure worked. After 102 trials for Sherman and 201 for Austin, they both could reliably name these three food items without error through thirty trials. Thus if I held up sweet potato, M&M's, or soycake, Sherman and Austin could easily select the appropriate lexigram, regardless of what happened afterward—even if I gave them nothing.

Had they really gotten the idea? Could they now use their symbols to indicate the names of things without expecting something specific to happen each time? The only way to tell was to show them a number of items not used during the fading training to see if they could generalize this concept to the other symbols in their vocabulary. They did so easily. We tried 21 additional trials with all new items, and they were both 100 percent correct.

It seemed like a giant bridge had been crossed. Sherman and Austin now really fully understood whether they were asking me for something or just telling me that a particular symbol was supposed to "go with" something. One was a communicative act, the other was merely a show of competence. Neither in itself was dramatic. What was important was that they could do both and they knew there was a difference. It meant that they recognized a distinction between communication for its own ends and communication about the properties of units of the system. I knew that linguists would dismiss it all as unimportant because syntax was not required and that some sort of "conditioning" explanation would be evoked by behaviorists. So I did not bother to write a paper saying that apes were able to understand the distinction between symbols as symbols and symbols as communicative devices—but I knew that a watershed had been crossed. Still, we were some distance away from something as simple as two-way communication between apes.

Before Sherman and Austin could really talk to each other, they had to become competent listeners as well as speakers. Other ape-language projects had understandably taken the process of listening for granted, assuming that if an ape could use a word, surely it could understand the word as well. I had already learned that it wasn't that simple. I had tried to get Sherman and Austin to ask each other for foods with the vocabulary they had. It was a disaster. Both chimps used symbols, but neither paid very close attention to the other, and certainly neither chimp picked up food and handed it to the other in response to a request—as I did when they asked me for food. Names they knew—but listening and cooperating seemed a world apart from the skills they currently possessed.

With a distinction between requesting and naming apparently established in their minds, we now had to move to receptive competence, or comprehension. This new skill would require that the chimps be good listeners, something no chimp had been taught in any ape-language study. I knew that teaching and testing for comprehension wasn't going to be easy, partly because of its "private" nature. But how could we expect Sherman and Austin to communicate with each other unless they knew what to do when they saw a word being used? Other apes like Lana and Washoe could respond to simple inquiries like "Are you hungry?" or "What is the name of this?" But learning to listen with your eyes and to cooperate when someone else says something like "Please pass the sweet potatoes, not the bananas" was an entirely different matter. It required more than just answering familiar questions with acceptable symbols or signs. It meant that they had to understand precisely what someone else was telling them to do, and they also had to be willing to actually do it. They could not get by just using words to indicate their own needs.

In our case the task was made even more challenging, because Sherman and Austin's vocabulary was small, thus limiting how we might interact. The simplest path would be to ask the chimps to give us something, such as a food item. Thirteen-month-old human children can perform such tests, but we knew that giving things—particularly food—was not among chimps' favorite activities.

My fear that Sherman and Austin would be reluctant to give me food items when I requested them proved unfounded, at least after a series of practice trials. But usually they failed to give the *specific* food I asked for. Instead, they gave their least favorite foods to me. Was decoding my request and searching for a specific food item too much to ask of the chimps? I needed another path to comprehension—one that would emphasize that in order to convey very specific information, I and others used words that they should try to comprehend and use to guide their ensuing actions.

I decided to use a hiding paradigm, in which my communications would function to reveal the nature of a specific food hidden in a container. The chimps already knew that when I entered a room with food items, they were allowed to ask for them using the keyboard. I decided that I would enter the room, after giving enthusiastic food barks to catch their attention and interest, but with a single food item hidden in a container. I could then use the keyboard to tell them what food I had hidden. I assumed they would want to know, so they could ask me for the food.

The first time I did this Sherman rushed to smell the container, but was unable to detect what was in it. He gestured for me to open the container, but I refused. Instead, I went to my keyboard, located just outside Sherman and Austin's room, and stated *this chow*. When I used my keyboard, it made the symbol "chow" appear on projectors located just above Sherman and Austin's keyboard. Sherman saw this information and apparently believed me because he immediately used his own keyboard to say *open chow*. On the next twenty trials of this novel situation, Sherman made just two errors, even though I used many different words. Sherman watched what I had to say each time and then asked for the food that I had indicated.

One of the errors he made was quite curious. On trial five, I told Sherman that M&M's, one of his favorite foods, were inside the container. Unlike the previous four trials, he did not ask for the food that I said was in the container. Instead, he asked for each of the four foods he had seen hidden in the container during the four previous trials! Did he think that I was simply going to repeat what I had done before and that he no longer needed to pay any attention? After trying four times to tell him that M&M's were in the container, and being ignored, I finally decided just to open it up and show him. His mouth fell open and he seemed somewhat astounded that indeed there were M&M's in the container, just as I had said. It seemed as though he could hardly believe his eyes. From that point on, he attended closely and seemed to believe everything I told him about hidden foods. I will never know, but perhaps Sherman was testing my veracity in saying I had M&M's.

Although the exchange, *this X . . . give X* might look like simple repetition on the chimp's part, I knew it was not. Known as match-to-sample, this simple sequence was one that Sherman and Austin had previously been incapable of engaging in. If shown one lexigram and asked to select its match, they inevitably failed such tests until they reached seven years of age.

If they could not match-to-sample, why could they then ask for the food that the teacher asserted was in the container? The communicative situation—made salient by desired food items—seemed to encourage comprehension of what was being said. It appeared that they understood I was telling them about the hidden M&M's. Once they realized that M&M's were in the container, they began to search their boards for this lexigram. It seemed that they were using the referential value of the lexigram to help them. That is, the symbol for M&M's on the projector made them think of M&M's hiding in the container. They knew what symbol to use to ask for M&M's and so could find it on their keyboard. Without a communicative situation like this, however, they could not match a symbol to itself.

My training as an experimental psychologist told me that this did not make any sense. Why could I not show Sherman and Austin the symbol for, say, banana and have them find it easily on their keyboard? Match-to-sample tasks were reportedly easy for Washoe, Lana, and Sarah. And why was it that when I *told* Sherman and Austin there was a banana hidden in a container, they could then readily find the banana symbol on their keyboard to ask me for it? If I analyzed only what they did, both were ostensibly reducible match-to-sample tasks. The principle of parsimony would require me to take the simplest possible interpretation—in other words, that they were matching a symbol to itself. Consequently they should have utilized this strategy regardless of how I interpreted their behaviors of interest in hidden objects and so forth.

Yet only if I took into account what *they* seemed to think they were doing did it make any sense that they could do one task and not another. Matching symbols to ones flashed on their projectors made no communicative sense, so they paid little attention to what they saw. However, learning about the nature of a hidden food made a lot of sense, so they watched closely and remembered what they saw. Yet there was no framework in psychological theory at the time that attributed any sort of thought process to animals. Nor was there any theory that even discussed the fact that the understanding of another's communicative intent might affect the way that the communication was perceived, or, indeed, if it was even perceived at all.

Many psychologists have steadfastly maintained that there are no important differences between a match-to-sample procedure and a communicative one such as we used—both are really just forms of elaborate imitation. After all, in both cases, the teacher says X, then the ape says X. Of course, whether or not these procedures differ depends upon the state of mind of the participants. If the chimpanzee assumes his job is to imitate the teacher, he will approach the situation with a different frame of mind than if he assumes that the teacher is conveying a privileged bit of knowledge about a hidden item. We could not directly ascertain the contents of Sherman and Austin's mind; nonetheless, it was clear that they assumed we were making a statement about the contents of the container.

The idea that chimpanzees could possibly approach communicative situations with any sort of hypothesis about the intentions of the communicator was not one that was widely accepted at this point in time. Consequently, my explanations regarding the importance of the manner in which the apes internally characterized the experimenter's goals met with general disbelief and occasionally outright anger. Sherman and Austin's expectancies, however, made all the difference. If, for example, I mistakenly informed them about the container's contents (which happened a few times), they would grab it from me and look persistently inside, as though they could not believe that I was wrong. Of course, they did not refuse to eat the food that *was* in the container—this would have been foolish—but they did persist in looking under it for the food that, according to my statement, should have been there.

Once Sherman and Austin could respond to information about the kind of food that had been hidden, we were ready to take the next step in building communication between them by using the system we had just developed. I wanted them to tell each other what was hidden in the container. After all, if they could now listen to what I had to say about the container's contents, perhaps they were ready to listen to each other. Moreover, since they knew the names of all the different foods I was using, they should have been able to tell each other what they had seen hidden as well, as long as one of them was permitted to see the container as the food was being placed inside.

On the first trial Austin accompanied me to the refrigerator, where he saw me put some banana slices in the container. My colleague Sarah Boysen, who had not seen the transaction, then took Austin to his keyboard and encouraged him to use the keyboard to identify the contents of the container. Austin quickly commented "banana," even though there were no bananas in front of him, only the closed container. Thus he had to recall the food that he had seen placed there. Meanwhile, Sherman had seen Austin go into the room with Sarah, with an evident interest in the container; and he saw Austin press the symbol for "banana." Sherman quickly went to his keyboard and also lit "banana." The chimps then shared their spoils. Again, this cannot be counted as simple delayed symbol-matching, because Sherman and Austin routinely failed at such tasks when we administered them in contexts devoid of communicative intent.

While we had no reason to believe Austin recognized on

the first trial that he was telling Sherman what was in the container, Sherman seemed to believe that Austin was describing the container's contents. After all, he had seen Austin go into the room to get the food, so it would be reasonable to assume that Austin knew something he did not about the container's contents. It is both interesting and important that no one had to show either Sherman or Austin that the chimp who saw the container being baited possessed knowledge that the other did not. The singular fact that the chimp who did not know what was in the container nearly always waited and watched until the other one revealed its contents was sufficient to show that the chimp recognized differing states of knowledge, based on observing, or not observing, the baiting process.

While the chimp who had not seen the bait seemed to know at once that he needed to watch what was said by the other, there did not exist an equally clear understanding on the part of the chimp who did know, that he needed to tell the other chimp. After all, if he knew, was it not sufficient for him simply to ask for the food for himself? Why should it matter if the other chimp saw what he said or not?

This was why we imposed the experimental constraint that required both chimps to ask for the correct food before the container was opened. Only if both had correctly requested the food in the container were the contents of the container shared. This meant that the "informer" needed to be certain that the other chimp paid attention and got the answer right. For the first eight to ten trials the informer (who had seen the food being placed in the container) simply requested the food for himself. He did not put himself in the mind of the uninformed chimp—not initially, at any rate. After both chimps had been the informer and the uninformed several times they began to realize that the informer had knowledge that the other chimp needed, if both were to receive food.

This understanding became obvious when one or the other of the chimps made a mistake. For instance, on one occasion Sherman (the informer) correctly indicated apple, but Austin hit the banana key. Perhaps he didn't believe Sherman, or was hoping that he would be able to get a banana, which he preferred.

When the container was opened to show Sherman and Austin its contents, it was clear who had made a mistake, and Austin tried to change his request to "apple." Thereafter, Sherman monitored Austin's behavior very closely, and if Austin looked hesitant he would urgently repeat his identification of the container's contents.

We could see, therefore, that through this procedure Sherman and Austin had learned two important features of humanlike communication: interindividual communication and cooperation. Not only had they learned *how* to communicate with each other, but they had also learned the *value* of communication—they achieved joint access to food items. Nevertheless, the presence of a teacher was important for the interchange to work. In the absence of the teacher, the chimps were unlikely to share the food voluntarily. In their natural habitat, chimps may collectively feed at a fruiting tree, but for the most part each individual eats alone, often turning its back on others. Only in the unusual circumstance of a captured monkey or other prey is food shared, and then in a rather grudging manner. If we were to establish humanlike communication between Sherman and Austin using food as a motivator, we would have to teach them to share food freely.

Initially, both chimps were extremely reluctant to share, as we had expected. One of us would sit between the chimps with a bowl of food, from which we would take a piece, break it in two, and give half to Austin and half to Sherman. They immediately turned their backs and ate the morsel in private, the way chimps do. Little by little—through a great deal of cajoling and patience on the part of the teachers—we broke down the basic chimpanzee tendency to avoid the sharing of plant food and replaced it with distinctly humanlike food dispersal behaviors. Although Austin was initially nervous about taking food out of Sherman's hand, because of the difference in dominance, eventually both he and Sherman shared food as humans would. When food was available, Austin would approach Sherman and expectantly request his share, even when no humans seemed to be present (we were watching out of view). Sherman would sometimes look away as if trying to ignore Austin, but Austin would persist and Sherman's face would begin to assume a guilty expression, as though he were aware of breaking a pact between them. Sherman would then hand over almost half of the food to Austin. The sharing was nearly always done quietly and calmly, and Sherman and Austin even began to look at each other while they ate.

It was an extraordinary sight, and many primatologists who visited the center were quite shocked to see such behavior. In a review article of our work, Carolyn Ristau and Donald Robbins, of Rockefeller University, suggested that we must somehow have coerced Sherman and Austin into food sharing, because it was so unnatural a behavior for chimpanzees. There was no coercion at all. Indeed, the chimps came to enjoy it as one of their most favored activities.

We were ready—we thought—to set up food sharing in a communicative context. After all, Sherman and Austin had the communicative skills of requesting and comprehending, and they were willing to share. We therefore set up a table with two trays of food, one for Austin and the other for Sherman. We then encouraged one chimp to request food from the other. They didn't. When Austin was making a request, Sherman was just as likely to be running playfully around the room as attending to Austin. Even when one of the chimps responded by picking up the requested food, he had to be encouraged by much gesturing on the teacher's part to fulfill the transaction. We realized that, left to themselves, Sherman and Austin were unlikely to use symbols for an orderly exercise of food sharing.

One of the key attributes of communication is simply paying attention to the other individual's actions—the role of being a good listener. This hadn't happened as Sherman and Austin sat around the food-sharing table (or rather ran around it), and so we decided to have them in separate rooms, joined by an observation window. Each chimp had a keyboard on which to make a request, which would show up on the responder's keyboard. With surprising ease, Sherman and Austin learned to attend to each other, attend to the requests made, and proffer the specified food. Being spatially separated in a rather formal way had facilitated the learning procedure for them. Having learned the benefits of sharing in this way, they later were able to transfer the process to the less formal arrangement of sitting around a table. They used their keyboards to make requests, and they responded with alacrity. Sherman and Austin frequently brought their food-sharing trays and table to us, looking expectantly. If we said, okay guys, we'll do some food sharing, they became wildly excited and ran around the room hooting, hugging, and tumbling in a general pandemonium of delight.

The food-sharing ritual began to be so much a part of their lives that they happily began also to take part in the preparation of the food table itself. This entailed carefully preparing two portions of each type of food and placing each portion group side by side in one of the compartments of the table. They eagerly vied to be the one who got to help prepare the table before each bout. Soon, food-sharing sessions could take place with no human present. Sherman and Austin simply sat themselves down and proceeded to have a meal, talking about which food they desired next and sharing all the portions. Occasionally they made innovations on the method of sharing that we had attempted to enculturate. For example, we had encouraged them to share each distinct food and made this easy by placing two portions of each food side by side on the food table. However, one chimp would sometimes eat both portions of a single food in response to a request. Then, given the next request, he would give both portions of that food to his partner. Of course, we cannot know what they thought as they innovated such alterations in the food-sharing concept, but we could not help but feel at times as if they were trying to say, "I know that I slighted you last time, but that was a food I could not resist . . . please accept two portions of this food in return." So much for coercion.

———

Sherman and Austin's ability spontaneously to communicate requests to each other with symbols advanced ape-language research in important ways. Not only was this the first documented interindividual communication, but it also incorporated degrees of comprehension not seen in other projects. Nevertheless, Sherman and Austin's vocabulary was small, simply because we had concentrated on other skills. In order to broaden communicative possibilities, we needed to build bigger vocabularies, including more food items and extending significantly into nonfood items.

We were delighted by the ease with which the chimps incorporated symbols for new food items, because it suggested that they were finally recognizing that each food had a unique correspondence to a particular symbol. To determine if this was so, I took a new food into the chimps' room along with their other foods as we were preparing to do food sharing. I carefully made no attempt to show them a new symbol for this new food. A number of unassigned symbols (or "unknown words") were present on the keyboard, though Sherman and Austin did not utilize them. Upon spying the new food, Sherman reached out for it and I suggested that he should ask for it. He quickly turned, scrutinized the keyboard, and deliberately, but with no fanfare, indicated one of the unassigned lexigrams as the symbol for the food. Austin watched attentively and subsequently used the symbol Sherman had selected to indicate that he would like to try the new food also. This unexpected skill encouraged us to believe that Sherman and Austin understood that a unique correspondence existed between each food item and a specific symbol. Moreover, they sometimes pointed back and forth between the symbol and the food item. They had named new foods and apparently agreed on coordinated use of the selected symbols.

When we moved onto nonfood items, however, we seemed to be back to square one. They learned to request a few objects such as blankets and keys, but when we asked the chimps to name these same objects they were unable to do so reliably—just as they had once been unable to name foods. Repeated practice produced little improvement, so we sought another way. When we attempted to introduce tools such as a wrench and lever, they began to confuse the tools and to string their tool symbols together, or to use the symbols interchangeably when requesting a specific tool. The challenge was to make object names more salient to the chimps—which essentially meant linking them to some activity that would make manifest the particular functional properties of each tool.

The procedure we developed was as follows. We built six food sites in the chimps' rooms, each of which required a specific tool to gain access to the site—key, money, stick, straw, sponge, and wrench. We put food in one site at a time and introduced one tool at a time. Ultimately, we randomly selected a tool site to be baited, then displayed the tray of tools and encouraged the chimps to survey the situation and decide which tool they needed and how to request it. Both chimps seemed fascinated by this task and were eager to work with the tools, even long after they were tired of using them to obtain food from the tool sites.

Although the chimps learned to request the tools with relative ease, two problems emerged as we proceeded. While constraining at the time, each problem provided important glimpses into the animals' minds. One problem appeared when we moved from the request task to the naming task. Although Austin very quickly began to learn to name each of the tools when they were held up to view, Sherman initially had great difficulty. He was able to request all six tools accurately, but for some reason he simply could not name them unless he intended to use them. It was a very frustrating time, both for Sherman and for me, as I would hold up a tool and he would select the wrong symbol. I would correct him, *No, this stick*, for example, and he would then respond *stick*. But even if I held up the stick the very next time he would answer incorrectly again. Sherman simply didn't seem to understand what was required of him.

This incomprehension continued for eighty-eight trials. I was then distracted by having to answer the telephone. On the eighty-ninth trial, with no change in procedure on my part, Sherman got it right. And out of the next 104 trials he made just four errors. It was apparent that he suddenly understood what I was asking him to do, and then did it. I've witnessed many times this kind of sudden recognition in chimps, and it has an uncanny human aspect to it—like the figurative light bulb going on in your head.

The second interesting problem we encountered involved functional errors with requesting and naming tools. Both Sherman and Austin experienced these problems, but in different ways. For instance, Austin frequently made wrench/key errors, using the symbol for one when the task required the other. Sherman had a syringe/key problem, often mixing up his request for them. In neither case was this because the symbols were similar for the pairs of tools. It seemed that the chimps were organizing their experience with tools in terms of their function. In other words, for Austin, the turning action of both key and wrench was most salient, while for Sherman, the insertion action of syringe and key was overriding. It was as if the chimps used the symbols as verbs, not nouns as we had intended. In any case, the different problems experienced by Sherman and Austin highlighted for me the individual nature of the chimpanzee mind.

Once Sherman and Austin had learned to request tools from a teacher, and to name tools reliably, we wanted to switch back again to communication between the chimps. We asked: What if only one chimp had access to tools, but none of the tool sites in his room had food in them? And what if the other chimp saw the sites in his room filled with food, but he had no access to tools? Would the chimps recognize that they could gain access to the food if they cooperated and communicated with each other? Would the chimp who had access to the tools watch the actions of the other at the keyboard, and willingly hand objects back and forth? Would they understand that they could make symbolic requests of one another for objects?

Sherman and Austin would need to do five things to achieve this: (1) attend to one another; (2) coordinate their communication; (3) exchange roles of tool-requester and tool-provider; (4) comprehend the function and intentionality of their communications; and (5) share their access to tools and the food obtained through tool use. If they achieved this, Sherman and Austin would have taken an important step toward using symbols much as humans do.

We approached this challenge using two rooms separated by a large window, as we had initially with the food-sharing task. On the first trial, all the tools were located in Austin's room. I then placed a food item in the wrench site in Sherman's room. Sherman immediately went to his keyboard and indicated *wrench*, but he gazed at me, not at Austin. I indicated I hadn't any tools or access to them, so Sherman then looked directly at Austin, who produced the requested tool. The chimps alternated roles as tool-requester and tool-provider, and very quickly seemed to understand what was required for the task: specifically, who had the tools and who had the information needed to select the appropriate one. The close attention to one another and to the series of actions that occurred was clear indication that the chimps comprehended what was going on and were not merely engaged in a mechanical chain of events.

As is often the case, we learned almost as much from Sherman and Austin's mistakes as from their successes. For example, on one trial Sherman mistakenly requested a key when a wrench was appropriate for the task, and he watched as Austin began to look over the toolkit in response to the request. Austin picked up the key, and Sherman looked surprised, turned to look at the keyboard, which still showed the *key* request he'd made, and realized his mistake. He rushed to the keyboard and corrected himself by tapping on the wrench symbol to draw Austin's attention to the changed request. Austin looked up, saw what Sherman was doing, dropped the key, and took the wrench to the window to give to Sherman. Such a sequence is indicative of intentionality and comprehension on the chimps' part, and cannot be dismissed as the rote result of conditioned response training.

Although these results were extremely pertinent to the ape-language debate, most researchers in the field paid little attention; the supposed primacy of syntax still held them in its thrall. There was, however, an assault from the behaviorist camp, specifically from Robert Epstein, Robert Lanza, and B. F. Skinner, the last being the most prominent figure in behaviorism at the time. In a 1980 issue of *Science,* they published a description of a choreographed sequence of behaviors in two pigeons, Jack and Jill. This sequence simulated communication that, superficially, looked similar to that achieved by Sherman and Austin.

Jack and Jill (both males) were housed in adjoining boxes with a transparent wall between them. Jack pecked a key labeled "what color?" Jill then looked through a curtain to the back of his box to see which of three colored lights was lit. He would then peck the appropriate one of three buttons, labeled R (for red), G (green), and Y (yellow); the pecking action illuminated the button. Jack then pecked a key labeled "thank you," which resulted in food being dispensed to Jill. Finally, after seeing which key Jill had illuminated, Jack pecked the equivalent one in his box, and was rewarded with a food item. The authors explained that, although the sequence of events might look like "sustained and natural conversation," it was in fact the result of strict conditioning procedures. They then went on to say that "A similar account may be given of the Rumbaugh procedure."[1]

It was at first humorous, but later somewhat frightening, to see an experiment of this nature given so much prominence. In reply, Duane and I wrote: "The description of this work seeks to parody ours and deceptively leads all but the most well-informed reader to conclude that, in fact, identical concepts were learned by chimpanzee and pigeon."[2] Whereas the pigeons had to be taught each step in the sequence and were not required to be aware of what the other pigeon was doing, the communicative behavior between Sherman and Austin emerged from components of true communication, and they achieved it on the first trial, with no training; they also had to be aware of their partner's actions and of the fact that they were indeed communicating with one another.

One issue that undoubtedly had provoked the behaviorists' attack was my conclusion that Sherman and Austin were exhibiting conscious intentionality during their communication—a clear red flag to those who believe behavior should simply be viewed as responses to external stimuli. As a result I became labeled a cognitive psychologist—one who believes that there is more to behavior than stimulus-response, and that animals may actually think about what they are doing. Sherman and Austin's behavior makes a strong argument for this view. I came to see clearly that intentionality must be accepted if we are to understand language in a

functional sense, and tracing the evolutionary emergence of conscious intentionality is crucial to the appearance and understanding of language as it is used by *Homo sapiens.*

We had taught Sherman and Austin key elements of communication—request, naming, and comprehension. Once these
were in place, other aspects of communication emerged spontaneously. The chimps began to pay close attention to each other's communications; they engaged each other before delivering
their message; they gestured to emphasize or clarify messages;
they took turns. None of these behaviors, all of which enhance
communication, was taught by us. Sherman and Austin developed them spontaneously.

Most important of all the communicative behaviors that
emerged spontaneously was that of indication, or announce-
ment of intended future action. At the start of one trial session
in which I was going to ask Sherman to give me objects I
requested, I apparently took too long to begin. To my surprise,
Sherman said *straw* and handed it to me. Then he said *blanket*
and handed that, too. Next he fingered the wrench, as if he
were thinking of it, then said *wrench* and pointed to the tool
while looking back and forth between me and the wrench,
checking to see if I had noticed the tool to which he was pointing. Initially, my impulse was to discourage Sherman, but soon I
realized the importance of the behavior. Announcing an
intended action is one of the earliest symbol-use skills to appear
in human children, and here it was emerging spontaneously
with Sherman. The same skill appeared in Austin, too, and it
wasn't always in trial situations. For instance, on one occasion
Austin announced at the keyboard he was going to make a
funny face, and then stuck his tongue out and pulled his lower
lip down over his chin. Comical though the context might have
been, it was linguistically important.

The more I observed such behaviors, as Sherman and
Austin sat at their tables independent of teachers, communicating their needs and sharing their food, the more their general
behavior took on a human countenance. This was not the result
of training but of use and development of communicative skills
in a social setting. It is true to say that we really didn't see lan-

guage competence in Sherman and Austin until food sharing fully developed, with its exchange of information and subsequent coordination of behavior. It wasn't a complex language, not a language with syntax. It was more a culture language, a complex set of behaviors that was the way the chimps' lives were lived in the laboratory. It made one think of *Homo sapiens* without sophisticated spoken language—intelligent, sensitive creatures, able to communicate and coordinate their behavior in a collective subsistence effort. I am not suggesting that what we have with Sherman and Austin is a precise model for a stage in human evolution. But it is easy to see how elements of social/communicative abilities, which are evidently present in our closest relative, the chimpanzee, could have brought about a new social/subsistence regime in our early human ancestors.

In light of the Sherman and Austin project, it is worth considering whether the probable parallel evolution in human prehistory of the skills of symbolization, tool-use, and interanimal communication was accidental. The answer, it seems, is that they are closely linked, within a complex social nexus. In such an evolutionary setting, selection for bigger brains would have been inevitable.

———

For me, the human countenance of Sherman and Austin's communication argued strongly that they understood what they were doing. I knew, however, that others would not be equally convinced by expressions. To most people, the wealth of information available from the ape's face is unreadable; to the unaccustomed eye, in fact, all apes look alike. Thus it was important to demonstrate beyond reasonable doubt not only that Sherman and Austin could communicate with each other, but also that they indeed knew that they were communicating. That is to say, we had to demonstrate that Sherman knew what he needed to tell Austin and why, and vice versa.

How could we prove that Sherman knew that Austin did not know which food had been hidden or which tool was needed? Recently, this issue has been given the somewhat confusing appellation, Theory of Mind. The idea behind this theory is that an important evolutionary breakthrough occurred when our ancestors began to realize that the knowledge that any one person had could differ from the knowledge that others had. Hence, minds were viewed as existing in different states of awareness. When the first individual became cognizant of this, it is said that she developed the theory that others had minds and that other minds differed from her own.

Did Sherman know that Austin's mind or knowledge state differed from his own? It seemed so; otherwise, why would he bother to tell Austin which tool he needed or which food was hidden in a container? As experimenters, we had set up the contingency that both chimps needed to ask for the correct food. But it was the chimps who began to attend to each other, to coordinate their communications, and to correct each other's errors. For me, as I watched them, the question was reversed: How could they not know that they were communicating? Yet I knew that most other people would not have the opportunity to watch them as I did and thus would remain skeptical simply out of lack of knowledge.

We could not simply ask Sherman if he knew that Austin's state of knowledge (or mind) could differ from his own. Consequently we decided to turn off the keyboard to see if they would attempt to communicate their privileged information in some other way. Every bit of training they had received to date revolved in one way or another around using the keyboard to communicate. We felt that if they attempted to communicate at all without it, we would have to conclude that they recognized that the reasons underlying the communicative process had to do with differing states of knowledge and the consequent need to share information.

So we baited a container of food with Peter Pan peanut butter while Austin watched and Sherman waited in the other room. We then took Austin and the container back to where Sherman was waiting in the adjacent room. Knowing that Austin had no sounds or sign at his disposal for communicating, we thought to leave in the room, on the floor, food labels taken off cans, boxes, and bags. The chimpanzees had never used these labels for communicating, nor had we ever purposefully set about to teach them these manufacturers' brand labels. Yet they always helped us prepare the food and thus saw the containers with these labels on them many times.

We thought that if they recognized these labels, perhaps they could think to use them as a means of communication if the keyboard were turned off. There were a number of very large ifs to be addressed in this situation. First, would Austin recognize that he needed to tell Sherman what was in the container that he, and he alone, had seen baited, even though the keyboard was turned off? And second, if Austin knew that he needed to communicate to Sherman about the peanut butter, would he recognize that the Peter Pan label, along with many other labels, was lying on the floor of the room? Third, if he recognized the label, would he think to show it to Sherman as a way of telling him the kind of food he had seen hidden? And fourth, if Austin did pick up the Peter Pan brand label and show it to Sherman, would Sherman recognize it and realize that Austin was trying to tell him that it was peanut butter in the container?

Much to our amazement, Austin and Sherman had no difficulty with this situation at all. When Austin found the keyboard was off, he immediately noticed the food labels scattered on the floor and then picked up the Peter Pan label and looked over at Sherman. Sherman stared at it intently for but a moment, then used his keyboard (which was on) to ask for peanut butter. Immediately after this, we took Sherman out to watch a container of food being baited with Welch's grape jelly and left Austin behind, thus reversing the situation. Sherman tried a bit harder than Austin to get the keyboard to work upon his return, but then, like Austin, he began to look around, noticed the Welch's grape jelly label, and picked it up and put it in his mouth, with the printed side down. He then proceeded to show it to Austin, not quite realizing that Austin could not see it because it was upside down. Austin did nothing. Sherman shrugged then put the label down, turning it over as he did so. Austin then saw what it said and immediately went to his keyboard and asked for jelly.

We continued to alternate trials in this manner through thirty different foods with manufacturers' labels, including M&M's, Jello pudding, Doritos corn chips, and so on. Austin and Sherman knew all thirty of these labels and used them essentially without error, from the very first trial forward, to tell each other the food that they had seen hidden in the container. They did this in spite of the fact that no one had ever taught them that it was possible to communicate using the brand names that manufacturers put on their food, and in spite of the fact that no one had ever taught them anything about communicating that did not involve a keyboard. Why would they have utilized a symbol system no one had taught them unless they recognized the need to share their differential knowledge? Even more intriguing was the question of how they had come to learn what each of these brand names represented, with no training at all. It had taken months for them to learn many of the lexigrams, and yet they knew all thirty of these manufacturers' symbols with no effort on our part at all. I began to wonder if we had not been using both the wrong symbol system and the wrong approach all along.

It was one thing to show that Sherman and Austin knew that they were communicating and that they even used a novel symbol system to do so. It was yet another to convince people that they indeed understood what it was that these symbols represented. Given my daily interactions with the chimps, it was becoming increasingly difficult for me to understand what it was that others were questioning and why. I recall a dinner in Washington at this stage of the work, in 1978. Most people there were psychologists, interested in animal behavior and human development. They were intrigued by our descriptions of Sherman and Austin, and seemed willing to listen to the suggestion that the chimps' behavior had gone beyond conditioned responses. "But," one of them insisted, "how can you *know* what they know? How can you *know* Sherman and Austin are using symbols referentially?" It was a good question, and one I needed to address experimentally. I spent the following day thinking how I might do this.

The issue of representation is fundamental for the field of

language acquisition, and we had to demonstrate that lexigrams were indeed representational to Sherman and Austin. I evolved the idea of the categorization test. Until this point in their experience, Sherman and Austin had used specific lexigrams for specific objects; that is, they had learned no generic names. If I were to be able to demonstrate that the chimps understood that symbols represented objects in an abstract manner, I needed to introduce the idea of generic symbols. We would first teach the chimps "tool" and "food" as generic terms, using just three items in each. group for the training regime. Ultimately, I wanted to see if Sherman and Austin, on being shown the symbol for, say, banana, could correctly categorize it by the symbol "food." The description of one symbol by another would, I thought, be strong evidence of the chimps' understanding of the referential nature of the symbols.

We approached the task in three stages. After first learning the category symbols "food" and "tool," the chimps would be shown food items and tools that had not been part of the training regime. For a successful test, Sherman and Austin would have to categorize correctly the new items on their first presentation. If a chimp had to be helped even once with one of the new items, that would count as training or imitation, and the test would be a failure. Our plan was then to repeat the process with plastic-covered photographs of food items and tools. Lastly, we would repeat it with symbols for food items and tools. This final test was the most critical of all, for it alone was unequivocally a test of the representational value of Austin and Sherman's symbols: It required a novel response, one never before given by the chimpanzee, or even the experimenter when working with the chimpanzee.

Apart from some difficulty Austin had in the second task in seeing clearly the food photographs within the plastic cover, both chimps scored successes on all three tasks. On the first trial of the final task, Sherman scored fifteen of sixteen symbols, and Austin seventeen of seventeen. For comparison, we performed these tests with Lana. Although she was able to sort food and tool items physically into groups, and therefore had some categorization skill, she was unable to do so using the generic sym-

Food label plaques presented to Sherman and Austin. The chimps had not been taught to recognize these logos, but were always interested in the packages as well as the foods. We constructed these "food plaques" by taping the package wrappers to pieces of lexan plastic.

bols "food" and "tool." This contrast between Lana and Sherman and Austin clearly shows that chimpanzees who have learned symbols have not necessarily learned the same thing. Because of their different experiences, Sherman and Austin had come to view symbols in a very different way from Lana.

The categorization study, I believed, showed as convincingly as anything can be demonstrated in behavioral research, that Sherman and Austin had developed a capacity of fundamental importance to language—the ability to use arbitrary symbols representationally. This, I felt, was sufficient justification for continuing ape-language research, despite the very negative atmosphere that had developed in the late 1970s and early 1980s. Such was the impact of Herbert Terrace's 1979 *Science* paper that most people inferred that he had declared ape language a dead issue. In fact, he did leave open the possibility that another research group, using different methods, might make important progress. The achievements with Sherman and

Austin represent that hoped-for progress. Unlike Nim, Terrace's chimpanzee, Sherman and Austin did not merely imitate their teachers: They were economic and informative with their symbol production; they did take turns in communication, both with each other and with their teachers; and finally, although much of their symbol product was request oriented, they moved beyond this stage to spontaneous symbol production either to announce or comment on their actions.

For example, one day while at the store, I noticed some unusually shaped colored plastic glasses. They looked like they would just fit in a young chimp's grip, so I bought them for Sherman and Austin. The next morning I showed them to the chimps and Sherman in particular seemed immediately interested in these glasses. This was unusual for him, since he generally could not care less what he ate or drank from as long as he approved of the contents. Austin was the one who generally focused on containers, often playing with them by stacking them or by pouring liquids back and forth from one to the other. Austin, however, seem uninterested in the new glasses. To my astonishment, Sherman carried his all around the lab, treating it like a prized and cherished possession the entire morning. At nap time, I put the glasses up. Later that afternoon, I returned with a strawberry drink in a pitcher. Forgetting how much Sherman had liked the new glasses, I grabbed what was handy, which turned out to be one old glass and one new glass. When Sherman saw me, he rushed to the keyboard to say *Glass strawberry drink* and then hurriedly pointed to the new glass to show me that he wanted his strawberry drink in the new glass, not the old one.

The important thing about this utterance was that neither he, nor I, nor anyone had ever used the symbol "glass" before. In fact, it was not an assigned symbol, just an extra lexigram on the keyboard that had not been assigned to anything. But Sherman was now assigning it to this favorite new drinking glass and pointing to the glass to let me know exactly what he meant and that he did not want his drink in the old glass. I glossed the lexigram as glass and began to use it for all glasses, but Sherman refused. He would use it only for these special glasses that he liked so much.

Once they were broken, which does not take long in a chimp lab, I could never find another and finally had to take the lexigram off the keyboard because I could not get Sherman to use it for anything else. I have never understood why Sherman liked those glasses so much. I purchased many glasses later on, some even looked similar, but none of the others seemed to matter to him.

Unfortunately, these unplanned demonstrations of cognitive capacity by Sherman and Austin were not considered "appropriate science." They were relegated to the realm of the "anecdote" by most who elected to follow Terrace in evoking syntax as the sine qua non of mind.

Herbert Terrace's primary interest was, of course, in sentences and syntax. He addressed the issue of the meaning of symbols in the following way: "What is important to recognize . . . is that neither the symbols nor the relationships between the symbols have specific meaning. Although the words and word order may be meaningful to a human listener, they may be meaningless to the animal producing them."[3] To Sherman and Austin, their lexigrams demonstrably have meaning.

How, then, do we answer our initial question: *In what sense* can a species other than *Homo sapiens* develop language? Sherman and Austin clearly have a language capability of a human sort, limited though it is in many ways. This conclusion is heresy to those linguists who would deny any kind of language capability to a species other than *Homo sapiens*. A second observation from the project is that the language capacity the chimps developed had to be cobbled together piecemeal; it did not emerge as a smooth developmental flow as it does in human children. This conclusion is heresy to linguistic theory in general. It also, I was to discover later, turns out to be wrong.

— 4 —

An Uncommon Ape

T HE bonobo (also called the "pygmy" chimpanzee) is a most unusual species of animal. The last ape to be identified as a species, the bonobo was continually confused with the chimpanzee until relatively recently. In fact, the bonobo is closely related to the chimpanzee. Around five million years ago, an evolutionary split caused a now extinct ape species to divide into two lineages: one that eventually led to *Homo sapiens,* and a second that led to modern chimpanzees and bonobos.

The chimpanzee is much more widely known, but it is the bonobo that is far more humanlike than other apes. In their anatomy, social behavior, vocalizations, sexual exploits, infant care, and mental abilities, bonobos possess an eerie human quality. Unfortunately, no one has yet tracked down the origin of the name "bonobo," and consequently many scientists maintain that they should be called pygmy chimpanzees. But they are not chimpanzees of any size or shape. They are more like persons with small brains and extra-long body hair.

In 1975 I began a Georgia State University postdoctoral fellowship at the Yerkes Center, where five bonobos were brought on lend-lease from Zaire. I therefore had the chance to observe the behavior of animals that had been wild born and reared. Observation of bonobos in the wild was still in its earliest stages, and so few were in captive exhibits that hardly anyone knew of their existence. Virtually nothing was understood about their behavior. Most zoological curators thought they were little more than geographical variants of the common chimp. Indeed, it was not uncommon for zoos to house them with common chimps or even gorillas.

My interest in bonobos began when I read the observations of Robert Yerkes, who wrote the book *Almost Human* in 1925 about a young ape named "Prince Chim." We now know Prince Chim was a bonobo, but at the time the species had not yet been "discovered." "Doubtless they are geniuses among the anthropoid apes," Yerkes wrote, after studying Chim and comparing him with a common chimpanzee, Panzee. "Prince Chim seems to have been an intellectual genius."[1] Yerkes described Chim as sanguine, venturesome, trustful, friendly, and energetic, while Panzee was distrustful, retiring, and lethargic. Chim's most human aspect, however, was the way he approached problems. "Most surprising and impressive in Chim's behavior was the continuity of attention, high degree of concentration on task, evident purposefulness of many, if not most, of his acts, and his systematic survey of problematic situations, his rapid elimination of unsuccessful acts or methods, and his occasional pauses for reflection," Yerkes wrote in a second book, *Chimpanzee Intelligence*. "I use this term [reflection] without apology, even to the behaviorist, for the simple reason that if Chim were a child instead of a chimpanzee we should apply the term without hesitation and with assurance that it would convey to every intelligent reader what is intended."[2]

Yerkes reported that both chimps displayed a range of humanlike facial expressions, but noted that this was particularly developed in Chim. Chim was also given to "pronounced laughter" as a way of expressing satisfaction and joy. On one occasion, Chim is said to have displayed his friendly nature by plucking some flowers and giving them to "a lady attendant." Both apes loved being out-of-doors, and they tumbled with abandon. "Frequently Chim would stretch out on his back in the pasture and with his hands under his head bask in the sunshine," wrote Yerkes. "It was strikingly suggestive of a human attitude of relaxation."[3] Apparently, Panzee never assumed such an attitude. "Never have I seen a man or beast take greater satisfaction in showing off than did little Chim. The contrast in intellectual qualities between him and [Panzee] may briefly, if not entirely adequately, be described by the term 'opposites.'"[4]

There are differences in personality among bonobos and common chimps, of course, so that Yerkes' comparison of Chim and Panzee may exaggerate the species' differences in general. Not all bonobos are intellectual geniuses, as Chim apparently was. Nor are all chimps dull-witted as Yerkes described Panzee. Overall, however, it is impossible to spend more than a few hours around the two species without being overwhelmed by the differences between them and the uncanny echoes of humanity one constantly experiences with the bonobo.

Yerkes, a professor of psychology at Yale University, had acquired Chim and Panzee from the New York Zoological Park and studied them for about a year. Chim died in July 1924, from pneumonia. Bonobos had not been recognized as a distinct species at this time. Reports of the existence of these unusual chimpanzees, living south of the River Congo (now the Zaire River), had only recently begun to reach the West, and few animals had been brought out. Scholars had known of the existence of apes since the seventeenth century, and the first scientific description of the chimpanzee was drafted in the late eighteenth century. For the existence of a species of ape to remain unknown for a further century and a half is a remarkable piece of scientific history. Consequently, the eventual discovery of the species was "one of the major faunistic events of the 20th century,"[5] as Dirk Thys van den Audenaerde, of the Tervuren Museum, Belgium, put it recently.

The first scholar to recognize the distinctive nature of the pygmy chimpanzee was Harold J. Coolidge, a zoologist at Harvard University. In 1926 and 1927 Coolidge was part of a university expedition to the eastern Belgium Congo (now Zaire), to collect gorilla material for the Museum of Comparative Zoology. The following year he visited European museums, including the Tervuren Museum, to gather yet more data on gorillas. Coolidge recently described his moment of discovery: "I shall never forget, late one afternoon in Tervuren, casually picking up from a storage tray what clearly looked like a juvenile chimp's skull from south of the Congo and finding, to my amazement, that the epiphyses were totally fused."[6] In other words, the skull, though small, was that of an adult. Coolidge found four additional skulls, all small, all adult. He made measurements and planned to write a scientific paper that would report the discovery of a previously undescribed form of chimpanzee.

Two weeks later the German anatomist Ernst Schwarz visited Tervuren and, prompted by Henri Schouteden, the museum's director, examined the same material that had so excited Coolidge. "In a flash Schwarz grabbed a pencil and paper, measured one small skull, wrote up a brief description, and named a new pygmy chimpanzee race: *Pan satyrus paniscus*," recalls Coolidge. "He asked Schouteden to have his brief account printed without delay in the *Revue Zoologique* of the Congo Museum. I had been taxonomically scooped."[7] For a zoologist, the naming of a new species is an important event, happening only once or twice in a career, if at all. Coolidge had apparently missed his chance, and so too had Schouteden, in whose charge the skulls had been for some years. In fact, Schwarz had thought the skulls represented only a subspecies, making it zoologically intimate with the common chimpanzee, which carried the subspecies name *Pan satyrus troglodytes*.

Coolidge continued his studies and in 1933 published a major scientific paper that raised what he called the "pygmy chimp" to the status of a full species, simply *Pan paniscus*. In that paper, he suggested that the pygmy chimpanzee "may approach more closely to the common ancestor of chimpanzees and man than does any living chimpanzee hitherto discovered and described."[8] This insightful comment lay unheeded for three decades.

In 1954 two German biologists, Edward Tratz and Heinz Heck, suggested that not only was the pygmy chimpanzee different enough to deserve the status of a full species, but that it was *so* different that it should be accorded a separate genus. The name they offered was *Bonobo paniscus*. They chose *Bonobo*, they explained, because it was the native word for "chimpanzee." In fact, no such word has yet been found to exist among the dialects of the people in Zaire, which is home to the pygmy chimpanzee. It has been suggested that the word may be a distortion of the town "Bolobo," a village where chimpanzee specimens had been collected in the 1920s. Tratz and Heck's call for separate genus status was not widely accepted, but the word bonobo has become commonly used as a synonym for pygmy chimpanzee.

———

The bonobo is not a true pygmoid form of the common species. While it is slightly smaller than the chimpanzee, weighing on average 84.5 percent of the larger species, the significant difference is in overall body shape. Bonobos have a more graceful build, relatively longer legs, and a smaller skull with a high forehead and an extremely expressive dark face. Because of their higher center of gravity, narrow chests, and more vertically mounted skulls, bonobos are able to walk bipedally more easily than the common chimp, a behavior that adds to their human-like aspect. In particular, when you see a bonobo walking on two legs, as humans do, you get a strong impression of what the human ancestor would have looked like, as Coolidge suggested.

This impression was cast in more scientific terms during the 1960s and 1970s when four researchers brought together three different lines of evidence in its support. From the evidence of molecular biology, the chimpanzee's anatomy, and what is known of the anatomy of the earliest members of the human family, the researchers concluded in a landmark paper in 1978 that "among living species, the pygmy chimpanzee offers us the best prototype of the prehominid ancestor."[9] The "prehominid ancestor" referred to would be a creature that predated the lineages that eventually led to modern chimpanzees and modern man.

Since Darwin's time, many creatures have been proffered as suitable "models" for the common ancestor, including the tarsier (a shrewlike primate), the monkey, the gibbon, and the African apes, particularly the chimpanzee. Yet there are potential traps awaiting those bold enough to attempt to reconstruct a missing piece of prehistory by looking to a living species as a model. One of these is the dangerous assumption that extinct species may be just like extant species. As Charles Darwin cautioned in his 1871 book, *The Descent of Man,* "We must not fall into the error of supposing that the early progenitor of the whole Simian Stock, including man, was identical with, or even closely resembled, any existing ape or monkey."[10]

By the 1970s, however, evidence from molecular biology clearly linked humans with African apes. Discoveries of human fossils, some more than three million years old, also displayed apelike aspects, particularly as regards the skull, face, and jaws. Such finds made it more plausible to look to African apes for clues to the anatomy and behavior of the common ancestor.

The earliest known species identified as a member of the "human family," *Australopithecus afarensis,* was essentially an ape that walked upright. Fossils dating back to more than three million years from Ethiopia and Tanzania reveal a head that was extremely apelike, with a small brain and protruding face. The most famous fossil is a partially complete skeleton of a three-foot-tall female, which was named Lucy by its discoverer, Donald Johanson. Although Lucy's gait was more upright, or bipedal, as indicated by the anatomy of her pelvis and legs, she displayed many apelike characteristics. For instance, in man the legs are much longer than the arms, while in apes the reverse is true. Lucy was intermediate between human and ape. The bones of her hands and feet were curved as well, suggesting that she spent a good deal of time in the trees climbing on branches. These apelike features suggested to many anthropologists that even though Lucy may have walked bipedally, she was, unlike ourselves, also a most adept tree climber. The popular impression of the earliest human progenitor is that of an ape-man striding bipedally onto the open savannah. However, it is almost certainly the case that this progenitor lived in a mosaic of woodland and forest and that it often had to climb trees as well as walk upon the ground.

Randall Susman, who has studied both wild populations of bonobos in Zaire and the anatomy of early hominids, notes that there are many adaptations for forest life in the skeletons of our progenitors. To the extent that the earliest human species were at least partial forest dwellers, it becomes even more reasonable to search for clues to our ancestor's behavior by studying modern forest-dwelling apes. The bonobo is the most forest adapted of the African apes, and is therefore a potential source of those clues to ancient behavior.

It is not just the habitat in common with *afarensis* that makes the bonobo a plausible model for the common ancestor. In 1982, Adrienne Zihlman, an anthropologist at the University of California, Santa Cruz, drew a now famous picture of a composite skeleton, with the right side that of a bonobo and the left side that of Lucy. The head, face, and jaw are strikingly similar between the bonobo and *afarensis*, except for somewhat smaller teeth in the human fossil. In the rest of the skeleton, the only evident difference is in the pelvis: The chimpanzee's is long, as an adaptation to quadrupedal locomotion, while that of *afarensis* is squat, as adapted to bipedalism. Aside from this, however, there is a close match between the two. The drawing clearly makes a powerful visual argument for the so-called pygmy chimpanzee hypothesis.

The hypothesis generated tremendous interest in the bonobo, and also provoked criticism. One criticism was that common chimps and bonobos are closer to each other genetically than they are to humans. The critics therefore asked: How could either species be a better model for an ancestral form? Zihlman responded in the following way: "The evolutionary question . . . is not whether one chimp is more closely related to us than the other is, for we know they are closer to each other than to humans. The question is, has one chimp species remained more similar to the ape-human *ancestor*, a 'living link,' while the common chimp and the hominids have undergone more morphological change?"[11]

In a more recent study, Henry McHenry, of the University of California, Davis, compared the anatomy of the common and pygmy chimpanzee, the gorilla, the orangutan, and *Australopithecus*. He confirmed some of the similarities between the bonobo and the human fossil, but pointed out that the bonobo did not have exclusive claim on the match. The shoulders, feet, and overall body proportions of the bonobo were the best match for the fossil human, but the common chimp is a better match for one of the arm bones (the ulna), the orangutan is better for the lower end of the thigh bone, and the gorilla for part of an arm bone. "The obvious conclusion from this is that the common ancestor of the African [apes and humans] was not precisely like any modern [ape] and its reconstruction must derive its form from clues provided by all extinct and extant [apes and humans]," concluded McHenry.[12]

McHenry's view becomes a more sophisticated version of the ancestral ape hypothesis, implying that we have lessons to learn about ourselves from all living apes. However, in the short time that bonobos have been studied, both in captivity and in the wild, it has become evident that Yerkes' impression of a human-like aspect to the species was correct. In this context, therefore, the bonobo is clearly the best choice among living apes as an ancestral model, and as such represents an important source of understanding about our prehistory.

Ironically, just as the world of science is becoming aware of the bonobo as a unique intellectual resource, the species is being pushed rapidly toward extinction. Bonobos live only in a small region of central Zaire bounded by the Zaire and Kasai Rivers. The area is rich in wildlife, including elephant, the okapi, the bongo, the forest buffalo, the duiker, and the l'Hoest and Hamlyn's monkeys. The humid forest that carpets the monotonously flat terrain is under pressure from several sources, including a rapidly expanding human population and commercial logging. Fifty percent or more of the bonobos have vanished in the past two decades due to hunting or deforestation. Bonobos are valued as a source of protein to people in some areas of their range, and as populations grow, so too does the need for protein. Furthermore, illegally captured and exported to foreign lands for a high price, the bonobos, chimpanzees, and gorillas represent a source of cash for local people whose resources are meager at best. Zaire has designated more of its country as protected parkland than any other nation, but the region where the bonobos live is not part of that system.

Two teams of researchers have been studying the chimps in their natural habitat since the early 1970s, one led by Randall

Susman, and the other by Takayoshi Kano, of Kyoto University, Japan. When Kano surveyed the region in 1973 he estimated the chimpanzee population to be about fifty thousand. "For six or seven years following the commencement of our investigations . . . , the forests of Wamba were a paradise for the pygmy chimpanzees and for us researchers," recalls Kano. "The pygmy chimpanzees, not bothered by the presence of humans, pursued a carefree life moving around the forest, and we followed them around without a care in the world."[13] Beginning in 1980, however, paradise turned into purgatory, as poachers and loggers began to plunder the forest. Instead of looking to the forest for survival, as man had done for centuries, African and non-African alike began to look upon it solely as an immediate source of monetary wealth. The fact that this wealth was clearly of limited supply only caused those wishing to take it to speed up the process.

Sensing impending disaster, Kano and his colleagues appealed to the Zairian government to set aside a region of the forest as a reserve. The appeal was granted in 1987, which put a halt to logging plans in the Wamba forest where Kano's research was located, and should have prevented hunting and capture. Poaching and killing continued, however, sometimes perpetrated even by government officials. Kano recalls an incident in 1988 when, in the absence of researchers at the station, regional officials entered Wamba and captured several bonobos for export. "That the formal establishment of a reserve was ineffective in checking the capture of bonobos, even by the regional government, came as a shock to me," he recalls. "The greatest disappointment was that some of the very people who were assigned the duty of protecting the (pygmy) chimpanzees, under the new law, had initiated their capture."[14] Many local people, increasingly including immigrants to the region, remained ignorant of the law, and hunting of the bonobo for food increased rather than decreased.

Now, two short decades since Kano's first visit, fewer than a quarter of the original number survive, at most no more than ten thousand. In 1990, the Bonobo Protection and Conservation Fund was established in the United States and

Japan, with the aim of protecting an area of six thousand square kilometers around the original Japanese research site, containing three thousand bonobos. It is a grass-roots attempt to salvage some small part of the the the last ape's former paradise. If nothing changes substantially, however, extinction of natural populations is inevitable, and the species will linger on for a while only in scattered captivity. With just eighty-five animals in zoos and primate centers around the world, that does not present a viable future. The last of the African apes to be discovered and the last to be studied scientifically in the field, the bonobo may be the first ape species to become extinct as a result of human activity.

———

The bonobos Lokelema, Matata, and Bosondjo, two females and one male, arrived at the Yerkes Regional Primate Research Center toward the end of 1975. They had been wild caught in Zaire, as part of an effort by the National Academy of Sciences to aid third-world nations by fostering the development of unique indigenous resources. These bonobos had left behind a familiar, lush forest and, after what was surely a nightmarish journey of change for them, now found themselves in a small cage, with dry food and surrounded by noisy, boisterous common chimps in neighboring cages. When I first saw them, they were afraid even to eat if anyone was watching.

My goal was to study their social and communicative behavior. I also planned to spend time observing a similarly composed small group of common chimpanzees housed in an adjacent cage. It was clear that whatever I learned about these bonobos would be partially constrained by the small size of the group and their artificial surroundings. But as virtually nothing was known about bonobo behavior at the time, I felt that it was important to begin with what was available. I also believed that these recently wild-caught individuals would provide a reliable index of bonobo temperament and ability. After all, if three New Yorkers were suddenly deposited in Zaire, they might not act as they would in New York, but they would nonetheless remain human beings. If they were well fed and healthy, it would be possible to observe a wide range of human behavior, regardless of whether they knew the ways of the jungle or not.

I was repeatedly told that I should not count on learning anything at all at Yerkes, because the bonobos were afraid of everything. Even observing was extremely difficult; if you were outdoors, they hid inside, and if you went inside, they hid outdoors.

My first task was to get close enough to these bonobos to be able to see them. This sounds strange for a captive setting, but it was nonetheless true. I decided that I would appear as if I, too, were nervous. I hid behind a brick wall near the cage and waited until they peeked out to determine if everyone had gone. I then hesitantly began to look around with wide eyes, just like theirs, but made certain I quickly retreated with a startled expression as soon as they spotted me. They were quite amazed that a human should be afraid of them, for humans had subdued them, captured them with nets, forced them into small boxes, and poked them with sharp objects with seeming abandon. But now, one was acting afraid of them. Slowly they became a bit bolder and would sit outdoors even when I was peeking at them. They knew that if they wanted to make me go away, all they had to do was flail a hand or stomp a foot in my general direction and I would beat a hasty retreat to the safety of my hiding place. By the end of the first week I was able to be seen quietly in front of their cage, and as long as I pretended to be interested in something other than them, they would stay outdoors and go about as though I were not there.

I then began to show an interest in what they were doing—not in them, but in their activities. For example, if they were playing with a pail of water, I watched every action with great fascination and looked longingly at the pail as though wishing I could play with it myself. I also showed them that I shared their fears and concerns by expressing similar ones myself, thus identifying with their view of the world. When caretakers came by who were talking and laughing loudly as they shoved food in the cage, I, like the bonobos, grew fearful at this loud and strange noise from the people who wore white suits and sprayed water on you and shot darts at you. I, too, shied away as these strange, boisterous, clanking men approached, not trusting any creature who so carelessly made noise and who always carried objects about with it, as if it were about to become angry and needed something ready to throw. After the first few weeks, I decided to go to the Yerkes kitchen and request permission to give the bonobos their daily rations and, much to my surprise, was granted it. Probably this was only because the caretakers noticed that the bonobos would actually approach me and take food directly from me, something they were all unable to achieve, except for James.

James, already long past retirement age, was the only caretaker still alive to have known Robert and Ada Yerkes. He worked with them in Orange Park, Florida, where the Yerkes center originated as an extension of Yale University.

James, as far as I could discern, was the only caretaker who could walk up and down the long rows of caged apes without being screamed at, spit upon, or plastered with the feces that always lay everywhere in the cages. When James looked at the apes, his eyes were full of care. Somehow the apes recognized this, even the ones that had become crazy from their isolation as infants.

The bonobos, too, had begun to trust James, and soon all of their care fell to James and myself. James began at times to speak to me, something that he rarely did with other scientists there, and told me stories of the "old days" when the Yerkes laboratory had been located at Orange Park, where it had been warm year-round and where the behavior of the chimps had been the focus of everyone's attention. I could see that James had been happy then, and that he missed those times, for things were now very different at Yerkes. He stayed on, it seemed, for the apes that he had come to love, and to try to show the new caretakers a few things—but most of them then only cared about drinking and carousing and seemed to view the apes as a nuisance.

In the 1970s, to work on the Yerkes great ape wing as a caretaker, you had to be male. I was permitted to be there only because I was a scientist. Had I applied for a job as a caretaker, I would never have been allowed on the wing. I was viewed as "odd" because I saw value in helping to scrub cages and because I was virtually the only female whom the apes ever saw. I was also the only scientist who spent any time with apes other than the minimal amount necessary for the collection of data. Other scientists thought I was eccentric in wanting to feed the animals or help in their care. However, I found it exceedingly practical, since this way their cages stayed clean and they came to trust me. Both factors made being around them a good deal more pleasant. I make these comments, not to set myself apart, but simply to describe how different were the worlds of those whose daily responsibility it was to care for and relate to the ape and those whose goal it was to understand them.

I knew from my experience with chimps at Oklahoma that groups of apes can easily develop an "us against them" attitude. I wanted Lokelema, Matata, and Bosondjo to feel I was part of "us" against the armada of white-suited caretakers and masked and medical scientists, veterinarians, and research technicians with their multiple needles, probes, charts, and rubber gloves. The idea that to approach apes you had to cover your entire body in white, from the top of the head to the feet, including the main medium you have for conveying messages to apes—the face—was appalling to me. Of course, the primary reason to cover up was to avoid the barrage of feces and water that were routinely tossed in your direction as you passed by the row of cages that formed the "wing." It seemed to me that it was much more appropriate to try to convey to the inhabitants of these cages that you meant no harm and that you wished only to pass, with their permission, of course. So I eschewed the covers and set about earning the right to pass unimpeded in front of the thirty cages of apes that separated the entrance area from the bonobos, who were housed at the very end of the "wing." This was important, for it was the only way I could reach the bonobos, and if all seventy-five apes between them and me were conveying messages of anger and aggression as I passed them, the bonobos would rightfully want little to do with me by the time I arrived in front of their cages.

Then there was the simple but obvious fact that you make a more appealing target in a white suit. So I also eschewed this accoutrement of science and walked about in shorts and a T-shirt. I was able to do so only because it was 1975, before strict guidelines about clothing that should be worn around apes were in effect. Now I would never be permitted to violate the white suit and mask code. Across months I slowly earned the right of passage and so, like James, I could walk past the cages unmolested. I regarded this as a great victory of sorts, for it signalled at least a modicum of acceptance by a very large number of apes. Almost everyone else at Yerkes regarded it as foolishness. Strange, I thought, that if your presence were accepted by these same creatures in the wild, it would have been regarded as an important achievement, but in captivity it was seen as foolish and bordered on skirting the rules a bit too much. Finally, I knew that at least Bosondjo and Matata, the juvenile bonobos, had accepted me. They sought me with their eyes at every unusual occurrence and began to identify with my reactions. They seemed to understand that I knew more than they about this strange human land they found themselves in. They also recognized that I was using my knowledge both to ease their way and to protect them whenever I was present. Soon it became clear that they trusted me enough that it would be possible to open the door of their cage and slip in, something I could never have done with wild-caught common chimps.

Throughout this time, I was enchanted by the ready ability of the bonobos to interpret body language and facial expressions accurately. I had known common chimpanzees with similar skills, though only those raised by human caretakers. Chimpanzees raised in the wild, or by their mothers in large social groups, seemed to have difficulty recognizing laughter, smiles, frowns, and many other human facial expressions. Bipedal stances were also threatening to them, as was a direct gaze in the eyes. Yet these wild-caught bonobos had no difficulty understanding the expression of these or other more complex emotions such as consternation, puzzlement, or gratitude, nor did I have difficulty seeing these emotions in their attitudes. We shared a language from the very beginning, albeit one that referenced mood and intent rather than specific objects. Had this communicative channel not been available, the tactic of relating my emotions to their concerns simply would not have worked. In the two weeks it had taken from first contact to entering the cage, I had therefore already learned that the non-verbal language of the bonobo was far more humanlike than that of common chimps.

The difference in temperament between the two species also quickly became apparent. The story of the plastic pails is a good illustration. From the very beginning, the bonobos had been given a plastic feeding pail in their cage, but it soon became much more than that. They used it for holding drinking water, inverted it as a seat, used it as a repository for urine, placed it over the head as a blind, carried it on the stomach as if it were an infant, played with it as a toy, and much more. The neighboring common chimps had been able to observe all these activities, and we wondered whether they would imitate the bonobos' antics if we gave them a pail. They didn't. Instead, they used the pails as props in aggressive displays, shaking them in the air, slamming them against the cage sides, and kicking them across the floor. Bosondjo was fascinated by these simian pyrotechnics. He picked up his own pail, carried it to a position in his cage from where he had a good view of the common chimps, and sat on it with his elbows on his knees and his chin on his hands, watching. When the display was over, Bosondjo got up, picked up the pail, shook it, and pushed it noisily across the floor, and then threw it at Matata. Bosondjo was obviously imitating what he had seen, but without the common chimps' intensity and aggression. He didn't do it again, and his pail remained intact for a long time. The common chimps quickly destroyed their pails.

I was helped in my study of the bonobos by Beverly Wilkerson and Roger Bakeman, and before long we were able to publish a series of papers that presented the first insights into the behavior of this extraordinary species. By this time, the mid- to late 1970s, biologists were very familiar with the social behavior of common chimpanzees, primarily through the pioneering efforts of Jane Goodall. Chimp society was seen to be

very much male dominated, with readily identifiable dominance rankings among both males and females. Each individual's social interaction was often directed to testing its status in
the dominance hierarchy. High-ranking individuals gained preferred access to food and, particularly important for males, to
mates.

Common chimpanzee females are receptive to sexual over-
tures only a few days each month, and during this time they evidence a large pink sexual swelling. Such females often find
themselves in the presence of a covey of interested suitors, all of
whom may mate with them repeatedly. However, Jane Goodall
and Carolyn Tutin have shown that multiple matings are not
the only strategy employed by male chimpanzees. They sometimes monopolize a female in estrus by insisting that she accompany them on a "safari" or a journey *en deux,* during which
other males are studiously avoided. This ensures the male singular access at the proper time.

Even with just Lokelema, Matata, and Bosondjo as our
small social group of bonobos to observe, it quickly became evident that bonobos in the wild had to have a very different
group structure and that sexuality functioned in a distinctly
bonobo manner. These early observations were later borne out
by the field studies of Takayoshi Kano, Suehisa Kuroda, Takeshi
Furuichi, and Nancy Thompson-Handler. Simply put, bonobos
are more egalitarian and utilize sexuality for a much wider range
of purposes than common chimps. Sex—both heterosexual and
homosexual—has expanded far beyond its initial function as a
means of reproduction among bonobos. It plays a central role
in bonobo society, as a tool for bonding all individuals together.

Bonobo life is centered around the offspring. Unlike what
happens among common chimps, all members of the bonobo
social group help with infant care and share food with infants. If
you are a bonobo infant, you can do no wrong. This high
regard for infants gives bonobo females a status that is not
shared by common chimpanzee females, who must bear the
burden of child care all alone. Bonobo females and their infants
form the core of the group, with males invited in to the extent
that they are cooperative and helpful. High-status males are

those that are accepted by the females, and male aggression directed toward females is rare even though males are considerably stronger.

Among both humans and bonobos, sexuality has evolved as a sort of multifaceted behavioral glue. If we disregard the aspect of genital contact for a moment and think of bonobo sexuality as "full body hugging," we can get a closer glimpse of the many facets of what we call "sex" in bonobos. Just as humans hug each other because they are happy, sad, or excited, and just as we hug others to comfort and reassure them, to greet them, to say good-bye, to show that we love them or are attracted to them, so do bonobos hug other bonobos in all such situations. Unlike ourselves, however, bonobos do not wear clothes. Consequently, when they engage in a full body hug, it is accompanied by some form of mutual genital contact, which is naturally stimulating.

Sexual arousal, excitation, and release in the bonobo is also a very rapid phenomenon, lasting about as long as most full body hugs in our own species. However, unlike humans, bonobos do not grasp the relationship inherent between sexual activity and reproduction. Therefore, they are freed from understanding anything other than the immediate consequences of their sexual expression. Humankind, on the other hand, has recognized this link and consequently cannot escape the implications of its sexuality. Offspring require extraordinarily long-term resources of time and energy. Activities that lead to offspring production will inherently, among humans, become subject to regulation and limitation of various forms because of the recognition that sexuality is linked to reproduction.

In the case of humans, not only is there a recognition of the link between sexual activity and offspring, but the human female is not physically equipped to provide for offspring without assistance. Human infants are much larger than ape infants and more difficult to carry as they do not cling. Consequently, a human female must be able to feed and defend herself while constantly supporting an infant who cannot manage to hang on while its mother engages in the activities necessary to sustain them both. For this reason, the assistance of human males, at least under natural conditions, is critical to survival of human offspring. Thus human sexuality not only leads to offspring, it also leads to the need to help the female rear those offspring if they are to survive. The realization of the implications of sexuality is a relatively recent product of the human intellect, and has not had time to become ingrained into the human biology. That is to say, we humans do not invent proscriptions upon our sexuality because we are genetically inclined to do so; rather, we invent such proscriptions because our intellect recognizes the need for their existence.

Freed from man's knowledge and responsibility for totally helpless infants, and equipped with a libido that is rapidly activated and released, bonobos are as free to engage in copulatory activity as a means of social expression as humans are to engage in full body hugging without mutual genital contact. Some primatologists have sought to "explain" bonobo sexuality by saying that it serves to reduce tension and defuse aggression. Such explanations beg the issue because they serve only to "explain" things that are themselves created by the conceptual framework in the mind of the observer. That is, once one assumes that tension exists, it therefore follows that it must be "released." Such a perspective inadvertently leads to the assumption that many behaviors, such as grooming or sex, exist for the purpose of "tension release." It is equally plausible to assume that the need to reproduce generates sexual activity and that sexual activity itself generates tension.

Bonobos, like humans, are concerned about who is copulating with whom, and sexual jealousy among and between sexes is present. Likewise, the advances of some individuals are refused, while those of others are accepted. These events produce among group members differences of feeling that require resolution. However, it is not the case that screaming and fighting typically precede sexual activity among bonobos. Bonobo sexuality, like human sexuality, is far too complex a phenomenon to be explained by the concept of "tension reduction."

Not only do bonobos utilize sexuality as a social glue, but they are keenly aware of their sexuality. Both sexes evidence interest in, and awareness of, the effect of sexual interactions on themselves and on their partners. Unlike common chimps, bonobo females initiate copulation just as frequently as males, and often with other females.

Also unlike common chimps, bonobos frequently copulate face to face, as humans do. Adding to their human aspect, bonobos peer intently into their partner's face prior to and during copulation, clearly monitoring change of expression. For instance, the intensity and frequency of thrusting is altered by changes in facial expression. Copulating pairs vocalize too, in clear communicative ways. We recorded at least three facial expressions and four vocalizations associated with copulation.

When Beverly Wilkerson and I reported our work in 1978 we said, "These observations strongly suggest that the pygmy chimpanzee is responsive not only to his own internal physiological feedback during copulation, but also to the subjective experiences of the partner."[15] I have no doubt that common chimpanzees are cognitive creatures, experiencing some degree of self-awareness and subjectivity. But this study of socio-sexual behavior was my first strong realization that subjectivity might be even keener in bonobos.

The existence of face-to-face copulation and prolonged eye contact would be sufficient to label bonobos as unique among nonhuman primates. But their sexual behavior has many other features that also set them apart. Most noticeable is the variability of copulatory partners and positions. Homosexual relationships between females are also a particularly obvious and prevalent aspect of bonobo life. Female-female friendships are cemented in this way and lead to a peaceful sharing of resources among females. Homosexual "copulation" between bonobo females (termed "GG rubbing" by primatologists) contains all of the components of heterosexual activity, except for intromission. Watching this behavior in Lokelema and Matata, I could see that the clitoris of the female became visibly engorged and erect and was rubbed vigorously against the genitalia of her companion. As in heterosexual copulation, the partners in GG rubbing gazed intently at each other and clearly derived pleasure and satisfaction from the activity. They also appeared to achieve climax, as evidenced by the uncontrolled rhythmic contractions that preceded termination of the GG rubbing. "The pygmy chimpanzee is the only one, out of 200 species of primate, to devise this behavior," said Kano when he later described GG rubbing he observed at Wamba, "and whenever they are delightedly absorbed in this 'lesbian' behavior, they seem proud of their splendid invention."[16]

Bonobos extend their sexual inventiveness further, to include male-male mounting, multiple positions in male-female copulation, and multi-individual heterosexual and homosexual copulatory bouts. Sexual activity is usually contagious in bonobo society; often, when one couple starts to copulate, the rest of the group members express interest in some way. In particular, GG rubbing between females seems to arouse sexual interest in males. Some individuals may initiate copulation with another partner, others may run close to the first couple, touch them, and scream in unison, or even join in with them. For instance, when Lokelema and Bosondjo copulated, Matata frequently joined in by rubbing her vulva against Bosondjo's back or mouth. In the larger group at the San Diego Zoo, a copulating couple frequently found themselves with as many as five young females vigorously rubbing their genitalia against various parts of the couple's bodies. Among bonobos, in sexuality there is an "interconnectedness of emotion and feeling" that goes considerably beyond any related behavior in other ape species—except perhaps ourselves. One cannot help but feel that at times they are simply seeking to share in each other's ecstasy, and that by so doing they heighten their own subjective experiences—much as many human cultures do during the nonsexual sharing of religious expression.

The more I observed copulatory bouts between bonobo couples, the more it became clear that the positions they assumed were not achieved passively. One or the other of the couple had a clear notion of the preferred position and used a series of gestures to indicate what was required. These gestures were of three types. First, an individual might place his or her hand on the partner's body, and move it in some deliberate way. This kind of positioning movement is probably the most primitive of communicative hand movements. The second type is more sophisticated, and involves a combination of touch and iconic hand motions, which are gestures indicating desired movement. The initiating individual lightly touches with one hand the part of the partner's body that is to move, and then with the other hand uses an iconic motion to indicate the nature of the desired movement. Last are completely iconic hand motions. Such gestures include moving the hand and forearm across the body, standing bipedally and waving the arms out from the body, and raising the arm with the palm down.

We used slow-motion videotape to study these gestures, and saw clearly that they were not randomly used. There was always a correspondence between the gesture of one individual and the subsequent movement of the partner, indicating a truly abstract communication system. Hand gestures might seem simple, but in fact they demand a high degree of cognitive sophistication. They require a clear concept of self and others. They also require the realization that personal desires can be communicated to another individual. Other requirements are that there be a temporary equivalence between the motion of the hand and the movement of the recipient's body, and that the hand not be acting as a hand in the instance of gesturing, but as a symbol for the recipient's body. These skills are often cited as important prerequisites of language. The iconic gestures probably evolved from touch gestures, and their efficacy evokes the sense of a primitive gestural language such as is hypothesized by some to be the beginnings of language in humans.

Bonobos turned out to be unusual not only for their outbursts of sexual activity when food was imminent, but also for the way the food was eaten. When I put food into the group of common chimpanzees I was studying, the largest, most dominant individual (Sonia) hoarded it and refused to share, despite aggressive displays, begging, and temper tantrums by her partners, Phineas and Little One. Only when Sonia was sated did she allow others access to the food. The bonobos' reaction was quite different. Sexual activity began just prior to the arrival of the food, followed by a frenzy of taking and sharing of food, and many instances of sex being offered in exchange for food. On some occasions Matata, for example, would take food out of Bosondjo's mouth in mid-copulation. Food sharing does occur among common chimpanzees, but only infrequently and then mostly between mother-infant pairs and under unusual circumstances, such as meat eating.

Despite the limited and inevitably artificial circumstances of our observations, we learned a lot. In the paper I published with Beverly Wilkerson in 1978, we said: "It is obvious at present, that behaviorally and morphologically, *P. paniscus* appears to be more like *H. s. sapiens* than any other living ape in every aspect of sexuality, from mutual gazing to homosexuality. The use of manual gestures to initiate copulation . . . lends support to this view."[17] We suggested that the socio-sexual patterns of the bonobo were an important social phenomenon that served to maintain unusual group stability. We also said that observations of wild populations, which were beginning to bear fruit at the time, would test this idea.

———

Although the existence of apes has been known for a very long time, systematic studies of natural populations began only recently, instigated by Louis Leakey. In the 1960s, Jane Goodall set up the Gombe Stream research site in Tanzania, to study chimpanzees; Diane Fossey established the Karisoke site in Rwanda, to observe gorillas; and Birute Galdikas went to Tanjung Puting National Park in Borneo to watch orangutans. Studies of the bonobo began a decade later, delayed in part by civil war. Two teams began work in the early 1970s: the Wamba forest team led by Takayoshi Kano, and the Lomako forest team led by Randall Susman. The region where the bonobos live, bounded in the north by the Zaire River and in the south by the Kasai River, is monotonously flat, carpeted with a mosaic of swamp forest and humid forest—and remote.

To reach the region from the United States, Susman and his colleagues must first fly to Brussels for a connection to Kinshasa, Zaire's capital city. From there to the Lomako forest requires an eight-day riverboat ride, a two-day trip by Land Rover, and a day's hike through swamp to the research station's base camp, which consists of several palm-leaf huts. The remoteness of the region explains the outside world's long ignorance of the bonobo, and until recently protected the animals from danger by human exploitation.

Making scientific field observations of primates requires a great deal of patience. Initially, you must habituate the animals to a human presence, just as I had done with Lokelema, Matata, and Bosondjo at Yerkes. In the forest, however, it is much more difficult; first, because the forest is dense and the animals are difficult to find, and second, because they can disappear rapidly into the canopy when frightened. The Japanese and American teams followed different practical strategies to habituate the groups of chimps they planned to study. Susman and his colleagues followed the "purer" route of patience, patience, and more patience, while Kano and his associates elected to entice the animals to a convenient location by provisioning them. They did this by clearing a small area of forest used by a bonobo group, and then occasionally putting out sugar cane, which the chimps love.

The method has the advantage of quickly getting animals to a location where you can observe them, but purists argue that the animals' behavior may be altered in subtle ways by the artificiality of the situation. As a result, Kano was able to collect more data, and more quickly, than Susman and his colleagues. Since there appears to be no significant difference in observed behavior between the Lomako and Wamba forest groups, Kano feels justified in his pragmatic approach. He points out that as long as the amount of food used is minimal, the apes treat the observation site as simply another place in the forest where they can locate food.

The social life of common chimpanzees, as mentioned earlier, is dominated by males. Social groups are relatively small, with perhaps thirty individuals, and loosely organized. Females with their offspring frequently forage alone, or in the company of a few other females, and over a limited range. Bands of mature males patrol the ranges of several females or groups of females, effectively herding them as their possessions and defending them from the attention of males from neighboring groups. The females generally leave their natal groups at adolescence, to spend the rest of their lives in another group. This means that male chimps in a social group are likely to be close kin—brothers and cousins. This kinship is a clear asset in establishing the strong bond that keeps the males together as an effective unit; they defend their females and procure new ones by raiding other groups. These intergroup clashes are extremely violent, and researchers at Gombe have recorded many bloody deaths as a result of them. For instance, one group split in two between 1970 and 1972, forming the Kasekela and Kahama groups. Over the next five years, six of the seven males in the Kahama group were attacked and killed one by one by Kasekela males.

Common chimp males are also aggressive within their own group, killing and cannibalizing infants on occasion. Infanticide of this sort often occurs when the mother is a recent transfer into a new group and her infant is likely to have been fathered by a male in her previous group. Such infanticide is seen by evolutionary biologists as benefiting the males in the new group by making the mother again receptive, this time to one of them. However, there are clear instances of infanticide among common chimpanzees in which the female had not recently transferred and the father of the infant was almost certainly one of the males participating in the cannibalization. Typically, these cannibalized infants are male and the taking of the infant from the mother seems to be preceded by generalized aggressive arousal among the males. Infanticide has also been reported among common chimpanzee females, and in such cases the benefits are more difficult to explain.

In all the years of observation at Wamba and Lomako, no example of murder or infanticide has been seen. This is despite the fact that bonobos live in much larger, more closely associated groups, sometimes with as many as a hundred individuals. In larger groups the opportunity for aggression is heightened, but bonobos have found a way to avoid it. The way, it seems, is to follow the 1960s dictum, "make love not war"—literally. The prevalent sexuality that I saw in the artificial settings of Yerkes and the San Diego Zoo operates just as much in the natural populations, and it acts to defuse potential aggression in all combinations of the sexes. "In pygmy chimpanzee society, the primary role of copulatory behavior is undoubtedly to enable male-female coexistence, not to conceive offspring," says Kano. "Here, copulation goes beyond reproduction."[18] The same may be said about homosexual behavior—that the primary role is social cohesion. As a result, says Kano, "the pygmy chimpanzee lives in a much more peaceful and mutually tolerant society than the common chimpanzee, its sibling species."[19]

The reason that bonobos often aggregate in such large groups is the nature of their food resources. Although their diet includes more than a hundred different species of food items, fruit is important among them. And the forest has many large trees that offer a huge food resource when they come into fruit. They are, however, spread apart in time and space. "Faced with scattered but concentrated food resources, pygmy chimpanzees improve their efficiency more by forming large parties than by separating into small parties," explains Kano. "To maintain a large size party, however, individuals must be able to coexist."[20] The high rate of sexuality among bonobos, with its humanlike aspect, is therefore likely to be the evolutionary product of adaptation to the particular nature and distribution of food resources.

Bonobo social groups are more closely associated and stable than those of common chimpanzees. Males and females occur in about equal numbers, as opposed to the imbalance in favor of females in common chimp groups. And, in general, bonobo males are less tightly bonded together than common chimps, and females more so. One reason for this is that, again, unlike the pattern in common chimpanzees, male bonobos remain closely affiliated with their mothers, long into adulthood. "In this way, the mother is the core of pygmy chimpanzee society, and the males lead a life following their mothers," observes Kano.[21] This same factor contributes to the much more equal status between male and female bonobos, in contrast with common chimps. Also, the closer bond among female bonobos allows them to form alliances to ward off any aggression from males. The overall more egalitarian tenor of bonobo society is reflected in the body size of males and females. In common chimps, where the males effectively compete with each other for access to females, males are about 15 to 20 percent bigger than females. (Such a size difference is common in species where males compete for females.) Male bonobos, by contrast, are only fractionally bigger than their potential mates.

Dominance hierarchies are not completely absent among bonobos, but they are more subtle and more fluid than in common chimps. "The breadth of variation in personality among pygmy chimpanzees is so great that a simple graphical representation of the dominant-subordinate relationships between individuals cannot be drawn," says Kano. "There is evidence, however, that pygmy chimpanzees do not have a strictly linear rank order."[22] The unusual influence of females in the society, combined with sometimes elaborate alliances between individuals, all too readily disrupts what fragile dominance hierarchy does exist.

Watching bonobo behavior in the wild is a beguiling experience, and it is easy to anthropomorphize, but the many humanlike qualities of these fascinating animals seems to demand it. Kano tells of many such experiences. When he was documenting the bonobos' food resources, he was surprised by the attention the chimps gave to earthworms. He often watched them dig diligently for hours in swamps, yielding a meager reward for their efforts. Sometimes, for instance, an efficient digger found a worm on average every twenty-five minutes. The bonobos at Wamba are not forced to forage at the margin of poor returns—they eat well there. It seems to be something of a leisure pursuit, observes Kano: "The activity resembles a household of people who leave for the coast on an occasional holiday, amusing themselves gathering shells at low tide and happily returning with a cupful of short-neck clams. In other words, it may be recreation. . . . It is hard to accept that the merit of this habit is the slippery sensation when the earthworm passes down the throat."[23]

One of the most intriguing observations at Wamba is branch dragging, an activity that involves technology and communication. Almost exclusively an activity of adult males, it appears to be a way of organizing the movement of the group on its day's activity. It therefore has elements of symbolism to it, much like the honeybees' dance. Sometimes an individual will grab a branch and drag it around boisterously, as an element of display. More often, and more interestingly, however, branch dragging is a prelude to group movement.

The individual often spends as much as half an hour searching for the "right" branch, frequently rejecting inferior ones. What criteria are being employed, no one knows. Then the chimp runs through the forest dragging the branch behind him, which creates a great noise and catches group members' attention. He might run back and forth several times, and he is sometimes accompanied by other males, also branch dragging. Finally, the entire group sets off into the forest—significantly, following the direction described by the branch dragging. Once the group is on the move, branch dragging stops. But if, for some reason, a decision is made suddenly to take a different direction from the one originally planned, the males leading the group will indulge in a burst of branch dragging, describing the new direction to be followed. Usually, a bonobo group on the move is relatively quiet. It is only when they reach a fruiting tree that a great vocal outburst erupts, an accompaniment to enthusiastic sexual activity.

Bonobos apparently sometimes employ branch dragging for personal reasons. Ellen Ingmanson, who has worked for some years at Wamba, once saw an incident in which Mon, an adult male, used the activity as a wake-up call for a friend. Mon had climbed out of his nest one morning and was sitting on the ground, looking up at another nesting tree and a fruit tree, about twenty-five feet away. After a while he got up, selected a sapling, broke it off, and began branch dragging between the nesting tree and the fruit tree. Soon, a head appeared from the nest as Ika, another male, looked to see what was causing the commotion. Mon stopped his branch dragging, squeaked excitedly, and jumped around. But Ika disappeared back into his nest. Mon resumed branch dragging, and after about five minutes Ika finally appeared again and climbed to the ground. Mon stopped once again and became very excited. The two chimps then set off for the fruit tree. "Mon had succeeded in getting a friend to join him for breakfast."[24]

Bonobos are often referred to as non-tool-users in the wild.

But branch dragging is clearly tool-use of sorts. Tool-use is a mechanism for solving problems. Common chimpanzees use sticks for extracting termites from mounds and stones for cracking nuts. That is readily recognized as tool-use. Pygmy chimps are blessed with plentiful food resources—they spend 25 percent less time foraging than do common chimpanzees. The problems they do face, however, are those concerned with controlling and manipulating social activity. Branch dragging constitutes tool-use in this respect, and indeed involves complex communication of a kind not thought possible in nonhuman primates.

The social skills that bonobos so excel at—including the exploitation of the power of sex—serve to make them the most successful of all primates, in respect of individual survival. Their devotion to reducing aggression in their midst optimizes bonobo individuals' chances of reaching adulthood. "They prove that individuals can coexist without relying on competition and dominant-subordinate rank," observes Kano.[25] How ironic it is that so peaceable and personally successful an ape should face extinction by the hand of *Homo sapiens*.

─ 5 ─

First Glimpse

K ANZI, a male bonobo, was born on 28 October 1980, and his entry into the world was as unusual as the things he would later enable us to learn about bonobos. His mother, Lorel, was on loan from the San Diego Zoo to the Yerkes Regional Primate Research Center, for breeding purposes. Because Lorel had been reared in the nursery rather than by her mother and because this was to be her first infant, I was concerned that she either might not know how to care for the infant, or might not wish to do so. I knew also that such problems could possibly be compounded by Lorel's low ranking in the bonobo group, and therefore suggested that it might be wise to isolate her for the birth. San Diego Zoo, however, decided that she should stay with the group as this was presumed to be a more natural situation.

Each evening toward the expected birth date, someone from the Georgia State University Language Research Center would drive to the field station at Lawrenceville, twenty-seven miles northeast of Atlanta, to be on hand to witness the birth and call the veterinary staff in case of complication. Rose Sevcik was on duty on the twenty-eighth, and she called me as soon as she was certain that the birth had started. By the time I arrived, Kanzi had been born, a tiny ball of black fur with spindly arms and legs. The situation with the bonobo group seemed a little tense. Instead of resting with her baby, Lorel was pacing about the cage looking tired and bewildered. The other bonobos were fascinated by the new infant and constantly pestered Lorel with requests to look at, prod, and hold the new baby. Finally, exhausted, she lay down and closed her eyes, with her arms around Kanzi as he clutched tightly to her waist, known as the ventrum, with his small hands.

Matata, one of the original three wild-caught bonobos in the Yerkes colony, approached Lorel and sat down quietly beside her, looking fondly at the new baby while cradling her own infant, Akili, in her lap. As Lorel peeked out of sleepy eyes, Matata gently caressed Kanzi's tiny hands, face, and feet. Lorel saw that Kanzi was not objecting and her eyes closed with heavy fatigue. Noting this, Matata leaned down next to Lorel and slowly slipped Kanzi's long, thin arms across her own ventrum. Lorel did not notice, so Matata carefully tugged on Kanzi's spindly leg and also slipped it onto her tummy. Kanzi's hand and leg reflexively gripped onto Matata. Lorel, at that point, seemed to nod her head as though passing from light sleep into a deeper stage and Matata, who is a keen observer, took account of this and used that moment quickly to pull the rest of Kanzi onto her ventrum. Kanzi, at once, clung to Matata, just as though she were his own mother. Lorel opened her eyes, looked at Matata, then down at her own ventrum and at once realized what had happened. She cried out, in a manner that could only be described as "plaintive" and tried to take Kanzi back from Matata. But Matata had already moved her own son onto her back and covered Kanzi so thoroughly by wrapping her legs and arms about him that Lorel could hardly see him at all. Lorel began following Matata around, tugging on her, trying to get Kanzi back. But she was hesitant to bite or attack Matata, perhaps because she feared that others would side with Matata, or perhaps because she did not want to chance hurting Kanzi. Matata was clearly determined to keep Kanzi, and Kanzi, less than thirty minutes old, had not had time to determine who his mother was, and so he clung to Matata and resisted the tugs and pulls of Lorel as well.

After several hours, Lorel gave up her attempts to pull Kanzi back and simply sat and watched Matata with her infant. Once Lorel was no longer trying to retrieve Kanzi, Matata began to hold him away from her body, as though she desired to look him over carefully. She held Kanzi about fifteen inches out from her ventrum and gazed intently at him as he vigorously moved his arms and legs, seeking to regain the comfort of her lap. Matata seemed to be somehow testing Kanzi, both by holding him out when he was clearly uncomfortable and whimpering, and by biting down gently but firmly on his fingers until he stopped clinging to her. It began to appear as though she were vacillating between "playing" with Kanzi as if he were an interesting toy, and caring for him, as if he were her own infant. Or perhaps she was simply doing this to frustrate and test Lorel. Lorel certainly became agitated with each of Kanzi's whimpers during these "inspections."

Just then the veterinarian, Dr. Brent Swenson, arrived. The bonobos fear the veterinarian as it is his job to anesthetize them, and he uses a blowgun to do so. This darting is an unpleasant experience for all concerned and the bonobos attempt both to avoid and threaten him. When Matata saw Brent, she tried to get as far away as she could and hid Kanzi in her lap to protect him. Thereafter she treated Kanzi as her own son, never again exhibiting the unusual inspecting and toying behavior she had engaged in just prior to the appearance of the veterinarian. She was a devoted mother to both Kanzi and Akili, nursing them jointly with skill and adroitness. Lorel continued to sit by Matata and paid close attention to Kanzi for several days, but after that she gave up and treated him as though he were Matata's infant.

Since Matata now had two infants, Yerkes offered to trade Akili for Kanzi and the San Diego Zoo agreed. They removed Akili and placed him in the San Diego group, while Matata and Kanzi traveled to the Language Research Center together. We kept Kanzi with his adopted mother to foster appropriate orientation toward his own species. Had Matata not "adopted" Kanzi in the way she had, the course of ape-language studies would have been entirely different, as Kanzi would have remained with Lorel and not been exposed to language during his infancy.

I had always been impressed by Matata's evident intelligence and eagerness to communicate. She developed ways of letting me know quite clearly what she wanted. For instance, she would hand me her food bowl and push me toward the refrigerator when she was hungry, or point to the lock on the door when she wanted to go out. Matata had been part of the language program for one year prior to giving birth to her son Akili. During that time, she proved to be a willing and interested, though incompetent, study. She quickly understood that Sherman and Austin used the keyboard to communicate and that pressing the lexigrams was what achieved this feat. However, the idea that a specific lexigram was used in specific ways eluded her. Matata would take my hand and lead me to the keyboard, fully cognizant of what she wished to convey. However, once she began to use the keyboard, she would press any lexigram and then look at me as though I should now know her wish. Consequently, her desires and the lexigrams she chose to use to express them did not correspond on a reliable basis. Sometimes she pressed "juice" when she really wanted a banana, other times she pressed "groom" when she really wanted to go outdoors. Often, she did not reject the juice or the grooming, even if her real intent was different, as there was no reason not to accept the juice and the grooming that I offered her. How then did I know that the symbols she selected did not match her real intent? Generally, either the nature of her glance, or the events that she was attending to gave her away. For example, if Sherman and Austin had asked to go outdoors and were getting ready to do so, Matata often wanted to go with them. She would vocalize and look in their direction, and if I permitted, she would rush over to them. Consequently, in such a situation, pressing "groom" did not really seem to indicate what she wanted to say.

Perhaps our hopes had been too high for Matata. Her early training had not been as systematic as that received by Sherman and Austin. She had not spent the long months learning to discriminate lexigrams, to request food items from a dispenser, to comprehend and carry out requests, or to differentiate between naming things and asking for them. Matata seemed so intelligent that we assumed she would be able to tell the lexigrams apart and utilize them for communicative ends.

Prior to this time, no bonobo had been language trained, and there were many reasons to suspect that Matata, as a bonobo, should do even better than Sherman and Austin. Robert Yerkes had observed that bonobos appeared to be intelligent and humanlike in many ways, and my earlier investigations of the gestural communicative skills of Matata, Lokelema, and Bosondjo had confirmed that.

Bonobos manifest a more intricate socio-communicative repertoire, including the use of more gestures and more vocalization, than common chimps do. These gestures are natural— that is, they are not trained but rather are reflective of the bonobo's own inclinations to communicate. In a paper with my colleagues Kelly McDonald, Rose Sevcik, William Hopkins, and Elizabeth Rubert, I expressed my expectations this way: "Because elaboration of the gestural, visual, and vocal domains of communication must have occurred in evolution before the emergence of speech proper, the more extensive development of these skills in the pygmy chimpanzee, in contrast to other apes, suggests that they might be better prepared to acquire language."[1] Matata was to be the test of that proposition.

———

Psychologists' understanding of language acquisition in children had advanced significantly by the late 1970s. Earlier, the simplistic concept of language had prevailed, with its extraordinary emphasis on syntax as the sine qua non of language. The newer concept, which was being developed by ape-language studies as well as detailed investigations of parent-infant patterns of interaction, was beginning to place language in a social context. Here, language is viewed as "a highly complex set of behaviors that is acquired through joint interactions that involve the intertwining of words and actions between two or more individuals."[2]

Both the results of the research with Sherman and Austin and new studies of children were suggesting that it was out of the combination of context, social interaction, and social expectancy that the child or ape steadily built up an understanding of the words that were being used around him or her. This understanding included the discovery that words could be used to refer to things, events, feelings, and so on. That is to say, words could serve as arbitrary replacements for objects and events in conveyance of meaning. This new account of language acquisition for the first time placed a major emphasis on comprehension of words and sentences as a key component of language skill in apes.

The first attempts at language training with apes had ignored comprehension, assuming that production of symbols implied comprehension. Our work with Sherman and Austin had demonstrated the fallacy of that assumption. It made clear that while apes *could* come to comprehend symbols, the skill had to be put in place through experiences designed specifically to foster understanding. Simply teaching an ape the association between a word and a thing did not always result in the capacity to comprehend the intentions of others when they utilized the same word. The associations between word and referent too often ran in only one direction; the ape declared what it wished to have happen and expected the proper events to follow. When the tables were turned, and someone else announced what they wanted to have happen, it was found that these associations were not always reversible. Thus much of the ape's understanding of language was limited to the ape as speaker.

Comprehending and producing language proved to be very different sorts of affairs. When apes produce symbols, they are attempting to affect the behavior of others—for example, to ask for a banana. When apes comprehend symbols directed toward themselves, they are expected to bring about the effect intended by the user of the symbols. Consequently, by focusing on the ability of Sherman and Austin to comprehend symbols, we were forced to develop paradigms in which the execution of the symbol and the ape's receipt of some object or activity associated with that symbol became completely detached. This marked a dramatic break with all other ape-language efforts, and it led to the apes recognizing that symbols can be used to communicate information about a specific object, event, or whatever without being tied to the occurrence of that event.

When Sherman and Austin reached this level of understanding of symbol use, we observed the spontaneous emergence of the capacity to use language to express future intentions. For the first time, it seemed that they really "had words"; that is, they understood that words could be used to express future intentions and thereby coordinate actions, rather than simply as a mechanism to get others to do something for them. Their ability to produce statements regarding their future intentions represented a profound advance in ape-language studies. Once apes could make statements about their intentions, and then carry out such statements appropriately, their behavior could no longer be explained by condition-response chains.

The work with Sherman and Austin had therefore set the stage for what we were to see in Kanzi in four ways: first, we had learned that "words" are more than simple associations between object and referent; second, we had learned that apes can appear to be able to produce relatively complex utterances without comprehending such utterances in the speech of others; third, we had discovered the importance of concentrating on language comprehension and the breaking down of the stimulus-response chains this entails; and fourth, we had found that comprehension leads to the ability to use language to make accurate and experimentally verifiable statements about intentions of future behavior.

We had thought that by working with a different species of ape we would gain further insights. Our initial results with Matata had not been encouraging, and Matata was fast approaching maturity. We hoped that perhaps, at some future time, we would be able to work with a young bonobo.

———

When Kanzi was six months old, this opportunity arrived. The Yerkes Great Ape Committee, which oversees the assignment of apes, assigned Kanzi to the Language Research Center as long as he could remain with Matata while he participated in language studies. I was looking forward to working with Matata again, because I had formed a close and trusting relationship with her. I knew that she would be excited to be at the new and spacious laboratory that we had recently been blessed with—fifty-five acres of primary forest on Georgia State University land. I anticipated that Matata would welcome a chance to be in the forest once again and that our relationship could be readily renewed. However, I was uncertain as to how Kanzi would react.

Unlike Matata, Kanzi had never interacted with me before, and all of his friends had been bonobos. Shortly after Matata and Kanzi were settled in their new quarters, Matata gestured to ask me to come into the cage. I entered rather circumspectly, hoping not to startle Kanzi. As soon as I was within a few feet of his mother, though, Kanzi emitted a piercing scream and leapt into my arms from Matata's ventrum, doing a midair acrobatic twist as he went. With his lips pulled back across both sides of his face, he looked directly at me, screaming with all the power his young six-month-old lungs could muster.

Having never been greeted by a bonobo infant before, I was uncertain how to interpret Kanzi's actions. Because his screaming was so intense, I worried for an instant that Matata might view this excitement as a sign that I was scaring Kanzi. Even though I counted Matata as a friend, I expected her to be fiercely protective of Kanzi, and to bite me if she thought my behavior potentially warranted any such action. I stood there, with Kanzi's arms and legs wrapped securely around my waist, trying to use all the nonverbal skills at my disposal to indicate to Matata that I was doing nothing bad to Kanzi in spite of the intense noise emanating from his small body.

I should not have worried. Matata understood Kanzi's emotion far better than I and thought it perfectly acceptable that Kanzi should scream and leap away from her to give me such a boisterous greeting. She continued to observe us as Kanzi gradually calmed down and began to explore my face, hair, and clothing. He was particularly fascinated by my nose, which he seemed to note was distinctly different from those he had been accustomed to seeing. This highly demonstrative greeting marked the beginning of what came to be a long and deep friendship between a human being who wanted to understand what it meant to be an ape and a bonobo who would strive with equal intensity in his own way to understand what it meant to be a human being.

The research plan laid out for Matata had been to pick up where her training had stopped several years back, when she was in our Yerkes laboratory with Sherman and Austin. We did not intend to work with Kanzi for some time, as six months was too young to begin any sort of systematic training. Consequently, Kanzi simply spent all day with Matata, doing whatever he could do to entertain himself.

Although Matata's previous progress had been disappointing, she had learned how to tell one lexigram from another, which was an important first step toward using them for communication. She could also make a few requests, but she still did not respond appropriately to usages by others. Thus she could not select an apple from the refrigerator if asked to, but she could herself ask for apples in a limited sense.

We began a systematic teaching program using essentially the same computerized keyboard equipment we had previously employed with Sherman and Austin, and we introduced new symbols to Matata very slowly, so as not to confuse her. Matata was eager to learn. She appeared to recognize that we were attempting to use the symbols for the purpose of communication. She was also patient, which proved quite a valuable trait, as Kanzi was a hyperactive infant. He ran around the test room, jumped on Matata's head, pushed her hand away from the keyboard as she tried to select the correct symbol, and stole the food she earned as a reward.

At times, Kanzi became mesmerized by the keyboard, staring at the symbols as they flashed onto the projectors at the top of the keyboard. He tried to grab each one just as it lit up, as though it was somehow crucial to catch it at just the moment it appeared. So occupied would he become with this activity that for ten or fifteen minutes nothing mattered to him except catching the symbols as they appeared. Then, he would turn and suddenly ignore them entirely, as if they did not exist.

Matata, like many other bonobo mothers, indulged Kanzi to the extreme. She made it plain that she did not approve if I or anyone else attempted to discipline Kanzi, regardless of the gravity of his mischievous behavior. She permitted him to behave in ways toward me that she would never consider doing herself. For example, if Kanzi was angry about anything that happened around, near, or with me or another person— regardless of whether we had anything to do with the situation—he was permitted to bite us. And we were supposed to ignore him, or better yet, distract him. It seemed that in Matata's view, Kanzi, as an infant, could not and should not be held responsible for his own actions—much as we, in our culture, also do not hold children or animals responsible for their actions.

When he was about fourteen months of age, Kanzi began occasionally to press keys on the keyboard and then run to the vending machine as though he had grasped the idea that hitting keys produced food. However, he gave no discernible indication of understanding the relationship between specific keys and specific foods. Rather, his use of the symbols was generally sporadic and playful. By the time he was two years old, he started deliberately to select the "chase" symbol. He would look over the board, touch this symbol, then glance about to see if I had noticed and whether I would agree to chase him. If I answered yes, either by smiling, nodding my head, or saying "yes" on the keyboard, he would run off, looking back with a big play grin on his face. He also began to use the "chase" gesture (a hand clap), which he had seen Sherman and Austin use between themselves. We were pleased with Kanzi's use of chase, but at that point did not recognize what this spontaneous learning portended for Kanzi's future.

Despite Matata's assiduous effort, and the systematic training program, she still did not progress as Sherman and Austin had. Able to communicate many things effectively through gestures, facial expressions, and vocalizations, she continued to be stumped by the keyboard. After two years of training and thirty thousand trials, she mastered only six symbols in a limited way. She could request things, but became confused when others asked her to respond to their communications. She remained tied to the anticipated consequences of symbol usage. When we attempted to alter these, by switching from a request task to a naming task or a statement task, for example, Matata no longer selected the correct symbols. She did, however, continue to try to convey things using the keyboard, albeit in an incomprehensible manner—something like the "word salad" that Terrace had described for Nim. She would become very frustrated when I did not understand her, as she pointed to a symbol that seemed to have little to do with where her attention was focused. At times she would take my hand, look me in the eyes, and vocalize urgently, as if trying to explain what she wanted and wondering why I didn't understand.

Matata differed from Sherman and Austin in other ways as well. For one thing, her ability to sort objects into recognizable categories was much less precise. She could sort only those objects that differed in many different dimensions, such as weight, color, shape, texture, and size. Objects that differed in only one way, such as color, were all grouped together. She was also unable to sort photographs—a favorite activity for Sherman and Austin. The inability to group photographs did not reflect an absence of picture-recognition skills, however, as looking at magazines was one of her favorite activities. She even sometimes "tasted" pictures of favorite fruits. By contrast, Sherman and Austin readily sorted objects that differed in one way, doing it first by color, then by shape, then by size, and so forth. They could sort photos into groups such as people, animals, tools, foods, and vehicles.

Tools also baffled Matata. She often tried to put a stick into a hole, with the stick oriented perpendicular to the opening. And she failed to learn how to insert a key into a lock, despite much practice. Many of her imitation skills were poor as well. For example, Sherman and Austin learned how to operate a joystick to control video games simply by observing me do so a few times. No amount of demonstration could convey the relationship between the joystick and the movement of the cursor on the screen to Matata. She finally learned, but only after a computer program designed specifically to shape this behavior in monkeys had been devised.

Why was it so difficult to repeat the steps into language taken by Sherman and Austin? Perhaps the problem was her age. At ten years of age when the project started, she was older than any other ape that had entered a language project. Could it be that apes, like humans, need to be exposed to language during infancy in order to acquire it?

Whatever the reason for Matata's difficulties, we had our own problems at this time. Our research grant was scheduled to terminate the following year, and if it were not renewed, our program of study would collapse. Matata's progress had been minimal and we could not blame the reviewers if they elected not to invest further funds in the study of bonobos and language. Not only had Matata failed to acquire symbols, we had learned little from her failure. Sherman and Austin's failures had always been informative, in that they had provided insights into the nature of language and how to proceed with further teaching. Matata's failure was leaving us empty-handed. We were unable to understand why she progressed so slowly.

———

While we were puzzling over Matata's poor progress, Yerkes decided that she should conceive again. They asked us to separate her from Kanzi and transfer her to the field station where she could be with Kanzi's father, Bosondjo. They also wanted to make the separation as easy as possible for Kanzi and they hoped to have Matata cycling at the time she joined Bosondjo. Consequently, they administered hormone tablets to inhibit lactation so that Kanzi would be weaned before she left and so that she would either be exhibiting a sexual swelling, or on the verge of doing so.

The plan was that Matata would go to the field station for a few months where she could breed with Kanzi's father Bosondjo, while Kanzi would stay at the Language Research Center. Kanzi would have to be weaned, and then a separation effected with the minimum of trauma. Prior to receiving hormone tablets, Matata had made no attempts to wean Kanzi. She permitted him to nurse at will. However, as the hormone tablets began to take effect, nursing became uncomfortable, and consequently Matata began attempting to stop Kanzi's nursing from time to time. At first he was puzzled and pointed at Matata's nipple while looking directly into her eyes, as if he were trying to tell her that he wanted to be permitted access to her nipple. After the incredulity wore off, a forlorn expression came over Kanzi's countenance and he attempted to get closer to his mother's nipple while pretending to be interested only in grooming. Finally, when it became clear that Matata knew what he was up to and simply was not going to let him nurse, Kanzi became furious and threw temper tantrums. Then, much to my amazement, he tried to enlist my help in this battle with his mother. He stretched out his hand toward me, gave a plaintive whimper, and with his glance directed my attention back toward his mother. He then began making threat barks at her. Having assumed, on the basis on my glances, that I agreed with him, Kanzi now threatened to bite his mother if she did not let him nurse. He was right to discern that I had felt sorry for him and was inadvertently taking his side in this psychological battle of wills. But I had no idea that he would use my support, conveyed only in glances and expressions, to escalate the conflict and threaten to bite his mother.

I was incredulous. Did he really think I would support him in attacking his mother and that I could make her let him nurse? Yes, I was sort of "in charge" of Matata from his perspective, but I certainly could not make her suckle Kanzi. It was a very curious position to find oneself in; being asked to intercede in a weaning battle between mother and son of another species. I decided quickly to withdraw my nonverbal and unintentional expressions of support. Kanzi then decided not to threaten his mother. He sat in her lap looking forlorn. I tried to help by making a bottle of milk and offering it to Kanzi but he refused it, preferring to sulk.

For several months before Matata's departure, Kanzi had been willing to travel with me to other parts of the center and around the forest, leaving Matata behind. At first, Matata protested loudly when she saw Kanzi go out of her sight, but eventually she was content for us to act as babysitters. Kanzi became intrigued by the "childside" of the Language Research Center, where children with mental retardation came for special language training. He seemed especially interested in the fact that the children used keyboards, just like his mother's.

By the time Kanzi was eighteen months old he began inventing simple iconic gestures, the first of which indicated the direction of travel in which he wished to be carried. He did this not with a finger point, but with an outstretched arm. Often, when he rode on my shoulders he added emphasis to his gesture by forcefully turning my head in the direction he wished to go—as though to guide my eyes. At other times, as he sat on my shoulders, he would lean his whole body in the desired direction of travel so that there was no mistaking his intent.

He employed another gesture to solicit help in opening a jar. This he did by making a twisting motion with one hand while pointing to the jar to be opened with the other one. When he wanted nuts cracked, he made hitting motions in the direction of the desired nuts. And if he wanted an object given to him, he gestured first at the person and then at the object, as human infants do. Often Kanzi vocalized while gesturing, which served to catch our attention and to convey the emotional affect that accompanied each request.

When the day of Matata's departure arrived, we indicated to Kanzi that we wanted to take him for a walk in the woods; of course, he was eager to go. While we were gone, Matata was sedated and placed in a van without Kanzi's knowledge. When Kanzi returned from his forest travels, he was not initially distressed at being unable to see Matata. However, after about thirty minutes of looking for her off and on, it became clear to him that she was not going to reappear just any moment. He began to appear anxious as his hair puffed out, and he started walking rather stiffly as his eyes assumed a worried expression. He seemed to think that his mother must be hiding somewhere, and he felt a need to look in every nook and cranny, indoors and out. He asked to visit every room in the lab, several times, as if he thought she might be moving about also looking for him. Kanzi then wanted to go back outside, and he searched many places there, too, including the trash cans.

That night he elected to sleep with Austin, but Austin didn't make the kind of big comfortable bed that Matata had, from intertwined blankets. After about an hour Kanzi joined me in a bed I had prepared nearby, in case he became frightened. For the next two and a half days, Kanzi glued himself to me and would not let me out of his sight. Finally, exhausted, I insisted that he stay with someone else. Kanzi objected. This was the first time I had heard Kanzi scream or express separation anxiety. I felt remorse over being the cause of any worry to him. Fortunately, he calmed down and began playing happily a few minutes later, so I used the opportunity to slip out when he wasn't looking, leaving him in my sister Liz's care. Liz had also helped raise Kanzi since his arrival at six months of age, and Kanzi knew her well. After searching briefly for me, Kanzi settled down happily in Liz's lap.

———

The day after Matata's departure, we set up the keyboard in the expectation that Kanzi would begin his language instruction—if he could learn to sit in one place long enough. Kanzi, however, had his own opinion about the keyboard and he began at once to make it evident by using it on more than 120 occasions that first day. I was hesitant to believe what I was seeing. Not only was Kanzi using the keyboard as a means of communicating, but he also knew what the symbols meant—in spite of the fact that his mother had never learned them. For example, one of the first things he did that morning was to activate "apple," then "chase." He then picked up an apple, looked at me, and ran away with a play grin on his face. Several times he hit food keys, and when I took him to the refrigerator, he selected those foods he'd indicated on the keyboard. Kanzi was using specific lexigrams to request and name items, *and* to announce his intention—all important symbol skills that we had not recognized Kanzi possessed.

How could this be? We had spent two years systematically trying to teach Matata a small number of symbols, with meager success. Kanzi appeared to know all the things we had attempted to teach Matata, yet we had not even been attending to him—other than to keep him entertained. Could he simply have picked up his understanding through social exposure, as children do? It seemed impossible.

For several weeks we monitored Kanzi's behavior and his use of the keyboard, checking to see if perhaps he was somehow fooling us into thinking that he knew how to use the symbols. We wanted to test his usage in a controlled situation, but because Kanzi had not been taught to do formal tests, he simply ran away every time we asked him to sit down and take a test. If we insisted, he cried and stayed there, but refused to participate. I consequently looked for ways to gather the data in a more casual way. We made tests into games that Kanzi enjoyed and that involved a great deal of play in exchange for a modicum of cooperation and data.

This was one of the most intriguing times of my professional life. I recognized that if what we thought we had accomplished proved indeed to be accurate, it could revolutionize our understanding of the nature of language acquisition, indeed perhaps of all learning processes. Equally significant, it could seriously undermine a main tenet in the body of knowledge around which both the social sciences and the physical sciences are constructed—that of the uniqueness of human mind. We would have to reformulate our view of apes as organisms. If apes could acquire language in the manner that humans do, without instruction, this meant that man did not possess a unique sort of intelligence, dramatically different from that of all animals. Perhaps *Homo sapiens* were given a gift for making speechlike sounds and the making of tools, but this did not mean that they understood things on a different plane from all other creatures. Kanzi's language acquisition seemed to announce dramatically that language acquisition was first and foremost a feat of understanding. The actual production of the sound was a matter of possessing the right peripheral apparatus to do so. The understanding of language, however, was a matter of comprehending the intended meaning behind the sounds, and Kanzi clearly was doing this.

For so revolutionary a scientific claim as this one, a persuasive body of data would be required, and as yet, I had only my notes of what Kanzi had done. Would anyone believe those? Would anyone believe anything without a number of detailed blind tests? I doubted it. I knew that convincing others would be a difficult task, but I also knew that if I were to focus too intently on proving everything Kanzi said or did, I would lose his natural engagement in the language process—a process that is comprised of communication and negotiation, not proof.

Given what Kanzi could already do, the only logical research strategy seemed to be to abandon any and all plans of teaching Kanzi and simply to offer him an environment that maximized the opportunity for him to learn as much as possible.

On a practical level, this meant rapidly increasing the number of symbols available to Kanzi, and figuring out how to make them interesting and important from his perspective. Since apes spend the better part of their day traveling from one food resource to another, I reasoned that the names of foods and locations should be among the most intrinsically interesting items to the ape mind. The Language Research Center's fifty-five-acre forest provided an ideal context for such learning, and the nature of the vocabulary would become pertinent if foods could reliably be found at specific locations as they are in the wild.

We set up seventeen locations in the forest where Kanzi could reliably find specific foods. He could travel between them as he wished. Different games and activities evolved at each location, thus helping make each unique. For example, at Look-Out Point, which was high off the ground, we often played a game of hiding objects. We hid pine cones in our shirts, pine needles under blankets, balls in the leaves, and so on. Kanzi loved these games of object hiding and would initiate them sometimes by hiding an object himself, and sometimes by saying things like *pine-needle hide* or *shirt hide*. At the location we called Tree-house, we rarely played hiding games with objects, but rather hid ourselves. Kanzi would sit high in the treehouse and wait while all of us hid ourselves in the dense foliage below. Then he would find us one by one. We always told him not to peek, but he hated that and would always look. Near the location known as Midway, Kanzi was always the one who hid. Here the thicket and undergrowth were dense and we hated to go into it. Kanzi would therefore use this opportunity to vanish quietly, usually no more than four to ten feet away, but nonetheless out of sight. We would call and call and pretend that we could not see him. He thought it great fun that he had eluded us and would sometimes sit quietly under a bush for ten or fifteen minutes before showing himself.

Travel was always accompanied by tree climbing, looking at small animals and insects, and learning about the naturally edible plants that were in the forest. We assigned words to these activities and objects and used them whenever the occasion warranted. No day was planned; it simply evolved, and each day was different. Sometimes we had to contend with dangers like thunderstorms, snakes, or floods; at other times there were visitors and surprises. Initially, either I or one of the other caretakers decided where to go, as Kanzi did not know how to use the symbols to initiate travel to a particular location.

On a philosophical level, the switch from structured training to laissez-faire learning meant that Kanzi, not an experimenter, would decide which words were acquired and what they meant. By believing I had to teach language to Sherman and Austin, I also had implicitly assumed that their abilities were limited and that we should decide which words they were ready to learn given our experimental questions. It now seems odd that anyone had ever decided which words an ape was to learn—certainly no one does that with children, who elect to learn very different sorts of words, apparently focusing on the aspects of language that fascinate them.

Could it be that our assumptions of limited abilities had inextricably led to circumscribed learning, even though we had been attempting to press the skills of Sherman and Austin to the limit? Could the same assumptions have limited the progress of Washoe or Sarah? If I had simply assumed that Kanzi was able to learn as humans do, might not those very assumptions have produced a very different sort of ape?

With Matata gone, I, and the other humans who helped care for Kanzi, became the focus of his life. He was now free to be fully attentive to the things we were interested in, without the constant maternal monitoring that had previously occupied him. Even though we wanted Kanzi to learn to communicate, we did not aspire to rear him as a human child. Instead, we wanted him to be a happy, well-adjusted bonobo who liked people. Nonetheless, we did encourage some humanlike activities, such as toilet training, simply because they made daily life more pleasant for everyone. Unnecessary human cultural predispositions, such as the need for privacy with regard to elimination, bathing, and so on, were not requested of Kanzi. Similarly, we did not attempt to clothe him for the sake of appearances or decency. Sometimes, when it was chilly out, he wore a sweat shirt or sweater, but he ruled out pants and shoes early on.

Kanzi was aware that we employed the keyboard as a means of communication and apparently felt keenly motivated to do so as well. He also felt motivated to cooperate with toilet training, though he had been indifferent to this activity while Matata was present.

Kanzi's communications soon began to revolve around his daily activities, such as where we were going to travel in the forest, what we would eat, the games we wanted to play, the toys Kanzi liked, the items we carried in our backpacks, television shows Kanzi liked to watch, and visits to Sherman and Austin. We found that the computerized keyboard was impractical for such outdoor use, and instead used a board on which photographs of the lexigrams were arrayed. Whenever Kanzi's caretakers talked among themselves or directly with him, they combined spoken English with pointing to the appropriate symbols. They also treated Kanzi's utterances as though they were intentional and conveyed what they appeared to convey, even if there was no "proof."

From Sherman and Austin, I had learned that the attribution of meaning comes from the behavior of the receiver of the utterance—and that the behavior of the receiver is critical to the maintenance of veridicality as words are being acquired. For example, when Sherman was learning the word "wrench," he often selected the symbol for "key," even though I knew he meant to say "wrench." Instead of telling him what he should have said, I acted as though he had intended to say "key" and responded by looking for the key and then giving it to him. In so doing, I let my behavior define for Sherman the meaning of his utterance. Thus the symbol "key" meant the object key to me, and that was what I focused my attention upon, even though I knew Sherman was wrong and that he needed to use another word. I did not tell him so directly. I let my behavior speak for itself to clarify for Sherman what "wrench" meant to me, thereby preserving the veridicality between the word and its meaning. If Sherman wanted me to act differently, for example, to pick up a wrench and give it to him, then he had to request that using different symbols—ones that "meant" wrench to me.

By maintaining a veridical response to each other's words, we establish a joint meaning for each word that is no more and no less than the sum of all such veridical responses within a community of common speakers. Therefore, the goal was not to determine whether Kanzi fully understood what a word meant when he said it, but rather to treat his saying it as a meaningful utterance and to respond as we would to any other party who used the word. Such responses, we had learned from Sherman and Austin, play a critical role in establishing "meaning."

Within four months, Kanzi's vocabulary rose from the original eight symbols to more than twenty. These symbols referred to foods, locations in the forest, activities, and people. By this time, too, he had learned the location of all seventeen food sites in the forest. He could announce his wish to go to any one of them, either by using the appropriate symbol or by pointing to a food that corresponded to a particular site (for example, at Tree-house there were always bananas and juice, at Crisscross Corners, blackberries, cheese, and orange juice were to be found, and so on); and he could guide us to any of the locations we named. After he had indicated his wish to go to, say, Tree-house, by pointing to a photograph of a banana, Kanzi might carry the photograph on the journey, frequently

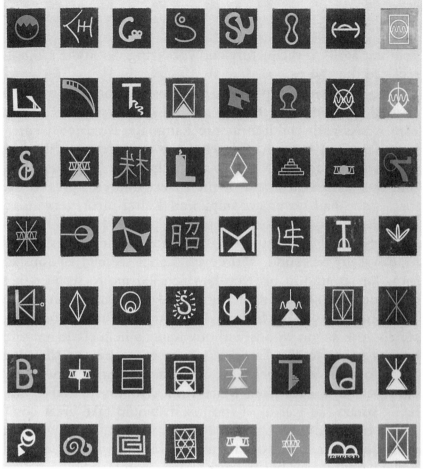

These symbols were added to Kanzi's keyboard at five years of age. Many are more abstract than some of his earlier symbols because we began to realize by that time that our expectations of him needing "concrete" symbols were mistaken.

The first symbol here is "bad," one that Kanzi acquired right away and often used to announce both his displeasure with us and his intent to do something that we were not likely to approve. Although Kanzi used "bad" only to refer to himself, his sister Panbanisha recently repeatedly used this symbol to comment upon Kanzi's action of biting someone she liked.

The next symbol is "now." We use this to tell Kanzi when we are ready to do something "right now" as opposed to "later" (also on the keyboard). Kanzi, however, rarely uses this keyboard symbol. He has devised a vocalization to communicate the same message and thus does not need the printed symbol.

The third symbol is "dessert," one of Kanzi's favorite things. He enjoys helping construct desserts and has come up with many of his own food mixtures.

The other symbols in the top row are "tummy," "bowl," "monster," "coconut," and "towel."

pointing again to the picture and vocalizing. Soon the symbols replaced the photos.

To test Kanzi's capacity to state his intentions to travel to various locations, we enlisted the aid of Mary Ann Romski, who works with children at the Language Research Center. Mary Ann and Kanzi had developed a close relationship during Kanzi's frequent visits to the "childside," but as Mary Ann was somewhat afraid of the snakes and creatures in the forest, she had never accompanied Kanzi into the woods. Mary Ann was therefore the perfect person for this experiment. She did not know the trails nor the locations where food was to be found. Moreover, Kanzi seemed anxious to show her around in his world, which led me to suspect that he would announce his intentions and lead Mary Ann to the locations he spoke of. I knew that Mary Ann would be completely lost in the woods without Kanzi's help. If Kanzi said he was going to the Tree-house, for example, and then led Mary Ann there, his behavior could not be explained on the basis of subtle cues from Mary Ann. Moreover, Mary Ann, being somewhat fearful of the forest, would take great comfort in being with Kanzi, a companion who was totally at ease in this element. She would trust him and his intuitions, and Kanzi would do his best to make Mary Ann feel "at home" in his world and to show her around.

Mary Ann recorded Kanzi's utterances and the route he led her along in going from one food site to another. During the test, Kanzi used photographs on five occasions and lexigrams on seven to announce a proposed destination, with 100 percent accuracy. Moreover, on all but one journey between two sites, Kanzi chose the most direct route possible. On that one occasion, he exploited the opportunity of being accompanied by someone who did not know that where they were to go was a part of the forest he normally is not allowed to go. Kanzi therefore demonstrated not only what he knew about himself, but also what he knew about Mary Ann.

Kanzi's capacity to use lexigrams to state his intentions was clearly evident in the test in the forest. Sometimes it took twenty minutes for him to lead Mary Ann to the destination he

Baby Kanzi shortly after he arrived at the Language Research Center. Kanzi arrived accompanied by his mother, Matata, when he was six months old. *(Photograph by Sue Savage-Rumbaugh)*

Kanzi at two years of age. *(Photograph by Sue Savage-Rumbaugh)*

Matata walks in the woods with baby Kanzi on her back. Foods were hidden under and around certain trees with markers on them. We found that we needed the markers to recall where we had hidden the food, but Matata did not. She could remember individual trees with uncanny precision. *(Photograph by Sue Savage-Rumbaugh)*

A bonobo in the wild carries her infant similarly. She must remember not only specific trees, but when they are fruiting as well. Bonobo mothers carry their infants up to five or six years of age, but only very young infants, less than five or six months, are carried continually. *(Photograph by Frans Lanting)*

By three years of age, it became apparent that Kanzi was learning lexigrams readily all on his own. He especially liked traveling out-of-doors, so we developed a portable keyboard. Here is our earliest attempt, a computer in a suitcase in 1983, long before the appearance of the first commercial portable computers. When Kanzi touched a symbol, it lighted up and the computer kept a record. *(Photograph by Elizabeth Pugh)*

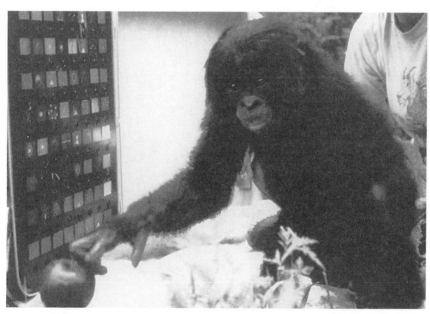

Kanzi uses the computer to comment "apple" after seeing someone take an apple out of the backpack. Kanzi often commented on things when he was small, whether he wanted them or not. *(Photograph by Elizabeth Pugh)*

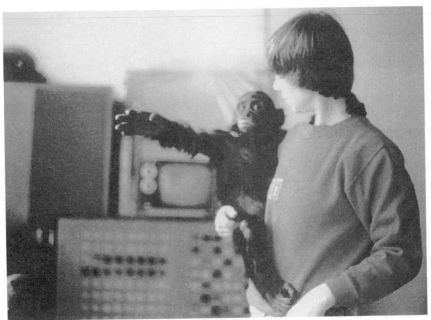

Kanzi shows me where he wishes to go by gesturing. He began to gesture at about twelve months of age and continued to use gestures intermingled with lexigrams after he began mastering the formal symbols. Bonobos do not typically direct others by such gestures in the wild. *(Photograph by Elizabeth Pugh)*

Kanzi at four years of age talking to himself on the keyboard. He began accompanying people on daily outings in the woods slightly before his third birthday. By four years of age he knew the forest far better than we did. He hated the cold months when he had to stay inside, so in the fall we let him wear sweaters to extend the time he could be outside. *(Photograph by Elizabeth Pugh)*

Between four and five years of age, Kanzi loved to take the keyboard aside and say things to himself. If we approached and tried to see what he was saying, he would pick up the keyboard and scurry further away. He could communicate by pointing to the symbols with the keyboard laid flat on the ground, but he preferred to prop it up in the vertical position as he saw us do. Here he struggles to properly stand his keyboard up. *(Photographs by Elizabeth Pugh)*

Kanzi watching one of his many favorite television programs. We set aside a scheduled time each day when Kanzi could watch TV. He selected his favorite programs by pointing to photos on the outside of the various tape cassette boxes. Often we made videotapes for him of things that were familiar from around the lab, using a "story line" to keep his attention. Kanzi typically prefers our "bonobo-oriented" videos, but also likes movies such as *Greystoke, Iceman,* and *Quest for Fire,* in which the actors portray "primitive man." *(Photograph by Sue Savage-Rumbaugh)*

Bonobos frequently express pleasure and happiness by smiling just as we do. Unfortunately, many people mistake Kanzi's smile for an expression of aggression because his teeth look frightful to them. Kanzi, however, uses exactly the same facial muscles to produce a smile as we do and his lips and eyes assume precisely the same countenance as ours when we smile. *(Photograph by Sue Savage-Rumbaugh)*

Lana, the first ape to use a keyboard system, constructs one of a number of complex "stock sentences" she was taught; here, "?You put piece of bread in machine." When Lana touched the symbols on the keyboard, the image was projected above the board in the order of her selection. After food was loaded into the machine at her request, Lana then had to extract each piece from the dispenser using other stock sentences, such as, "Please machine give piece of bread." *(Photograph by Frank Kiernan)*

A computer "read" Lana's sentences, and if they were grammatically correct and corresponded to the food that was actually in the machine, the computer vended Lana a piece of the requested food. Lana revealed to us that she could go beyond the "stock" sentences to construct novel appropriate requests. For example, when the machine was broken and food could not be loaded, Lana was able to ask, "?You move food into room?" *(Photograph by Frank Kiernan)*

Austin (on the right) has just watched as a particular type of food was hidden in the plastic container. Austin has learned that he can have this food only if both he and Sherman ask for it correctly. Sherman, however, does not know what food to request, since he did not see the food as it was placed in the container. After realizing that Sherman does not know what to ask for, Austin uses the keyboard to identify the food he saw hidden. Sherman watches, then asks for the correct food. *(Photograph by Sue Savage-Rumbaugh)*

Austin and Sherman sharing food at a picnic using their version of the portable computer keyboard. By ages eight and nine, they had mastered ninety-two symbols and could readily use them to communicate with each other as well as with people. *(Photograph by Elizabeth Pugh)*

One day Kanzi wanted to go for a ride in the lab van but found it locked. When no one would give him a key, he tried the next best thing—a screwdriver that the shop foreman had left nearby. *(Photograph by Elizabeth Pugh)*

Kanzi's younger sister, Panbanisha (one year old here), studies a photograph of one of the places where food is hidden in the forest. When the apes were too young to understand symbols, we often showed them photos of the places we planned to go in this way. They liked this practice immensely and studied the pictures intently, as Panbanisha does here. *(Photograph by Sue Savage-Rumbaugh)*

Feral born and reared, Matata studies her mirror image. Although she has not been able to learn language, like most apes she quickly recognized herself when given the opportunity to gaze into a mirror. Matata uses the mirror to facilitate the extraction of tiny hairs from her chin, so she will not appear to have the least sign of a "beard." We do not know where she acquired this practice, because it was not demonstrated for her; perhaps bonobos have something akin to vanity in the wild. *(Photograph by Elizabeth Pugh)*

In this important experiment, Sherman has been shown some bananas being put into a box on his side of a divided room with a window separating him and Austin. Austin does not see this happening. The box has been locked, so that Sherman will need a key to open it. Here Sherman goes to the keyboard and states "give key." *(Photographs by Sue Savage-Rumbaugh)*

Austin watches from his side of the room and reaches into the tool kit for the key.

Sherman (left) approaches and Austin hands him the key.

Sherman goes back to the box, inserts the key into the padlock, and twists the key to open the lock.

(Above left) Austin watches carefully as Sherman opens the lock.
(Above right) Sherman (left) takes half of the food back to Austin (he tastes Austin's portion on the way).

Sherman passes Austin's portion to him through the opening in the window.

This is one of the indoor rooms where Kanzi now plays and works when he cannot go outside. It includes a television with an attached joystick and two keyboards mounted in study racks. On the floor is a bin of objects and photographs for Kanzi. *(Photograph by Steve Winters)*

Austin watches a video-tape of himself. *(Photograph by Elizabeth Pugh)*

Working with Kanzi at the keyboard. *(Photograph by Steve Winters)*

In one of the early blind test formats presented to Kanzi, experimenter A (with back turned) places three photographs inside a gray plastic "text booklet" and then passes the closest booklet to experimenter B (facing Kanzi). Experimenter A then plays a tape-recorded word to Kanzi. The word is randomly chosen from Kanzi's suspected vocabulary. When Kanzi hears the word, he gesturally asks experimenter B to open the test booklet and show him the photographs. Kanzi looks over the options, selecting the one that corresponds to the word he heard on his headphones. After Kanzi makes his selection, experimenter B tells Kanzi whether or not he is correct. In this example, Kanzi has selected a photograph of fire, and when experimenter B turns over the answer card, she finds the word *fire*. *(Photographs by Nicholas Nichols)*

Kanzi often enjoys drawing and painting. Sometimes he labels his drawings, but as yet he has not achieved sufficient skill to produce recognizable images—a skill that also eludes most children under the age of three. *(Photograph by Steve Winters)*

Kanzi plays grab, tickle, and chase with Jeannine's son, Nathaniel, who has learned lexigrams himself. Yerkes veterinarians do not permit apes to interact directly with children for fear of transmission of childhood diseases, so they are separated here by a plastic barrier. *(Photograph by Nicholas Nichols)*

Kanzi bashes one stone against another to produce a small sharp-edged flake which he will use to cut through a thick nylon rope holding tight a cover on a box filled with a favorite food. *(Photograph by Nicholas Nichols)*

This panel represents one quarter of the symbols on Kanzi's keyboard. By the time Kanzi was seven years of age, he comprehended three-quarters of the symbols on his 256-symbol keyboard, though he regularly used only about half of them. Symbols he did not learn were very abstract, such as "away," or for things that he rarely encountered, such as pomegranate. From left to right, the symbols stand for: Bad, Now, Dessert, Tummy, Bowl, Monster, Coconut, Towel, Taco, Chicken, Lettuce, Noodles, Sugar, Bunny, Burrito, Butter, Away, Slap, Salt, Observation room, Trash, Perrier, Bottom, Strawberry, Kiwi, Pillow, Pomegranate, Grapes, Privet berries, Panzee, Pinkey, Yogurt, River, Jello, Lana, Backpack, Koolaide, Wipie, Noise, Popsicle, Swimming pool, Panbanisha, Cold, Draw, Middle test room, Carry, Shop, Thank you, Karen, Honeysuckle, Toy, Dan, Gorilla, unassigned, Book, unassigned. *(Photograph © Language Research Center, Georgia State University)*

Here Kanzi and I are at Lookout Point in November when it is starting to get fairly chilly. Kanzi has just told me he wants to go on to Flatrock to look for M&Ms and for a ball. *(Photograph by Nicholas Nichols)*

had earlier specified, and along the way he often talked of other things, without forgetting where he was going.

———

Matata's breeding sojourn was successful and we were pleased to find that she would soon return to the laboratory, pregnant. Despite the joy that another bonobo would bring, however, we wondered if Kanzi would continue to use the keyboard when Matata returned. After all, he had rarely used it before she left. Maybe he would simply want to stay with her again and the importance of his "human friends" would fade into the background. We talked endlessly about what should be done. Some suggested that we should not let Kanzi rejoin Matata, for fear we would never learn what he might be capable of communicating. Keeping a young bonobo away from his mother, when both lived in the same laboratory, seemed to me to be an impossible feat, as well as an unbearable cruelty for the sake of science. We reached a compromise in which Kanzi would be allowed to interact with Matata through the bars of a cage for the first few days. We would then evaluate how we should proceed, based on Kanzi's interest in continuing to communicate.

Matata arrived at the center one day in September, drugged, while Kanzi was in the forest. I stayed with her in the colony room as she came out of the anesthesia. She was delighted to see me, and vocalized happily when she heard Sherman and Austin's calls from nearby. Eventually, Kanzi returned, hot and tired after a long day in the forest. I sat down with him and told him there was a *surprise* in the colony room. He began to vocalize in the way he does when expecting a favored food—"eeeh . . . eeeh . . . eeeh." I said, *No food surprise. Matata surprise; Matata in colony room.* He looked stunned, stared at me intently, and then ran to the colony room door, gesturing urgently for me to open it. When mother and son saw each other, they emitted earsplitting shrieks of excitement and joy and rushed to the wire that separated them. They both pushed their hands through the wire, to touch the other as best they could.

Witnessing this display of emotion, I hadn't the heart to keep them apart any longer, and opened the connecting door. Kanzi leapt into Matata's arms, and they screamed and hugged for fully five minutes, and then stepped back to gaze at each other in happiness. They then played like children, laughing all the time as only bonobos can. The laughter of a bonobo sounds like the laughter of someone who has laughed so hard that he has run out of air but can't stop laughing anyway. Eventually, exhausted, Kanzi and Matata quieted down and began tenderly grooming each other. How could anyone have contemplated not allowing them to be together? If we've lost him back to the world of bonobos, so be it, I thought to myself.

Just then, Kanzi gestured for the keyboard and indicated *open*. I opened the cage door and went in. Kanzi climbed on my shoulders and gestured for me to go through the door. I was shocked. Surely he will change his mind as soon as we leave Matata, I thought. But he didn't. We got Matata some bananas and juice, she vocalized with contentment, and Kanzi looked at me and said, *childside*. Off we went, to see who was still there at the end of the day, something Kanzi often liked to do. All our concern had been for nothing. Kanzi had elected to be with Matata and also with *Homo sapiens*; he was to negotiate two worlds.

For seventeen months we kept a complete record of Kanzi's utterances, either directly on the computer when he was indoors, or manually while outdoors. By the end of the period, Kanzi had a vocabulary of about fifty symbols. Within a month of separation from Matata, he was already producing combinations of words, and he continued to do so throughout the seventeen-month period of our first report. Other apes had also produced combinations, but Kanzi's were different in that they reflected a competence born of comprehension, rather than a need to form longer symbol groups to answer questions, as had been the case with other apes. Nearly all (more than 90 percent) of Kanzi's multi-word utterances were spontaneous; that is, they were not responses to teachers' requests or imitations of teachers' utterances. By contrast, Terrace found exactly the opposite pattern in his chimpanzee, Nim. Three-quarters of Nim's multiword utterances followed something a teacher said. Spontaneity of utterances, rather than their frequency, is a critical aspect of language use, for it is the spontaneous utterances that tell us something about what another party is thinking. Utterances that are prodded out of an ape in response to questions such as *Who Nim hug*—while Nim is already hugging someone named Laura—do not really provide the listener with new information, meaning information that is not already self-evident in the situation.

Kanzi, however, formed spontaneous utterances such as *Matata grouproom tickle* to ask that his mother be permitted to join in a game of tickle in the group room. This happened on one occasion shortly after Matata had vocalized to him. Since Matata typically did not join us in the group-room play sessions, this request on Kanzi's part completely surprised us. Perhaps Kanzi knew from sounds that Matata was making that she wished to join us and thus sought to tell us about her wishes. We cannot know for certain; however, we do know that we did not prod this utterance from Kanzi, nor would we have considered bringing Matata into the group room were it not for Kanzi's communication. Many of Kanzi's multiword utterances had this character of novelty and functioned to suggest completely new actions and alternatives to our normal ways of doing things.

It was also the case that as Kanzi added more elements to his utterances, the information content increased. For instance, one of Kanzi's multiword combinations was *ice water go* (with *go* indicated by a gesture), by which he was asking someone to get some ice water for him. A combination such as *play me Nim play* is typical of Nim's utterances, which contain a great deal of redundancy. Nim may have been more loquacious than Kanzi, but what Kanzi was producing was more like language.

Of Kanzi's three-item utterances, the most interesting— and significant—were those in which he indicated someone other than himself as the agent or recipient of an action. Most of his three-item combinations involved the initiation of play, such as grab, chase, and tickle. Some of these games involved

Kanzi directly, but others were intended for his teachers. For instance, Kanzi might indicate *grab chase* at the keyboard, and then take one person's hand and push it toward a second person: the chaser and the chasee. Statements of this sort were Kanzi's inventions, as none of us suggested we play with each other, leaving Kanzi as spectator. Compared with food requests or requests to be tickled, where the chimpanzee is always the recipient of the action, statements initiating action between two other individuals is complex. As my colleagues and I observed in a paper describing these utterances: "Clearly, prior to the emergence of syntax must be the emergence of the concept that one can request that A act on B, where the speaker is neither A nor B."[3] The fact that Kanzi mastered this betrays a real sense of self and of others.

We tested Kanzi's language competence in many ways throughout the seventeen-month period, and one significant feature that emerged was that his comprehension of symbols consistently preceded his use of them. This is the human pattern of language acquisition, and the opposite of how we had concluded from our previous work that apes learned. Moreover, we formed a steadily growing conviction that Kanzi also had a good comprehension of spoken words. This is a difficult issue, one fraught with strongly held beliefs and little scientific data. Many investigators have assumed that apes easily come to understand English, a view held by Charles Darwin. Referring to other animals, too, he wrote in 1871: "[T]hat which distinguishes man from the lower animals is not the understanding of articulate sounds, for, as every one knows, dogs understand many words and sentences."[4]

So strong is the impression that Sherman and Austin can comprehend spoken English—and, perhaps too, so strong is the urge to believe it—that visitors to the Language Research Center have often refused to accept our explanation of how they appear to do so. Even when we point out that careful tests, which eliminate contextual information, have demonstrated they do not really have the level of comprehension that appearances suggest, these visitors still choose to believe the chimps can understand the spoken word.

Apes are so aware of, and competent in interpreting nonverbal aspects of, communication that they often infer a speaker's intent while not truly understanding the words. If you watch a soap opera on television with the sound turned off, you will gain a sense of the kind and breadth of information that can be conveyed without language proper—it is rather extensive. The ability to "read" information into the situation from a variety of sources, including gestures, glances, actions, intonation, and knowledge of similar previous situations, is highly developed in apes. It is this capacity to read the "meaning" inherent in a given situation that often convinces people that apes are understanding language, for we humans focus so exclusively on language, once it is learned, that we are often unaware of the other channels through which we are gaining information.

When we conducted tests on Austin and Sherman that eliminated contextual information, forcing them to respond to the auditory stimulus alone, their answers were clearly random guesses. In these tests, Sherman and Austin were presented with words that they knew well as lexigrams and were asked to select a matching photo whenever they heard the word. They did well whenever they saw a lexigram, but not when they heard the speech. We could therefore be certain that they understood what they were being asked to do. Often, when they heard only a word, they even gestured to the keyboard, asking me to "say" the word in lexigrams so that they could find the correct picture.

These findings are in accord with observations made of chimps reared in home environments earlier this century, by Louise and Winthrop Kellogg in the 1930s and Catherine and Keith Hayes in the 1940s. The chimps, Viki and Gua, displayed limited comprehension under test conditions, and those words they did understand were closely tied to a specific context. The Kelloggs spent many hours teaching Viki the names of various body parts, but with no success, and the Hayeses drilled Gua with the names of familiar objects, again without success.

More recently, Roger Fouts has suggested that Washoe, Ali, and a few other apes understand at least three or four words. Unfortunately the data are few and the number of overall correct responses, even to a word that is thought to be understood, are extremely low. Even though the data suggest that there is a statistically significant difference between words that are recognizable and those that are not, these words are understood on such a small percentage of occasions that most oral communications made in real time would be missed. Francine Patterson also suggests comprehension of speech for Koko the gorilla, but she lacks the test data to rule out the impact of contextual information. It may be that other apes show some speech comprehension, but none of them seem to have made the leap into language comprehension that Kanzi was on the verge of at this time.

—

We first detected what seemed like spoken word comprehension when Kanzi was one and a half years old. We began to notice that often, when we talked about lights, Kanzi would run to the switch on the wall and flip it on and off. Later he simply looked at the light switch on hearing the word, apparently visually forming the pairing. How was this different from what we and others had seen in other apes? Was Kanzi not responding to some undetermined contextual cues? At first we thought this must be the case. However, there were differences. Sherman and Austin responded to speech and contextual information when someone was talking to them or saying something of relevance to or about them. They often responded appropriately to queries like, "Do you want to go outdoors?" "Please don't push the TV set," and "Stop tickling so hard." Kanzi also responded in this manner, but in addition, he seemed to be "listening in" on conversations that had nothing to do with him. For example, I once asked another caretaker if "someone had left the lights on last night" and then noted that Kanzi was gazing at the light switch, even though neither I nor anyone else had done so, and in fact we had not even been discussing the particular light that he was noting.

It was what seemed to be this uncanny ability to "listen in" on conversations that were not directed to him, and indeed ones that he was not supposed to pay attention to, that first began to convince me that Kanzi was processing sounds in a manner that I had not experienced in other apes—even in those who had been reared in human homes. I believed, however, that this was not possible. The prevailing theory of the time was that of Phillip Lieberman, who asserted that vocal and auditory capacities co-evolved in all creatures. According to Lieberman, the auditory system, composed of the inner ear and the auditory cortex, was capable of processing sounds in an efficient manner only if they corresponded to the types of signals that a species made with its vocal tract. An animal's auditory comprehension was, so to speak, tuned to the capacity of its throat by co-evolution of the communicative process. This theory seemed very reasonable and it explained, for me, why Sherman and Austin had difficulty understanding words even though we spoke to them constantly.

As time passed, Kanzi appeared able to understand more and more spoken words. Finally, I could no longer ignore the fact that what Kanzi was doing really did not fit within the standard theory and that it was not like the kind of apparent comprehension or partial comprehension that I had encountered in Sherman and Austin or Lucy and Washoe. Kanzi had become so proficient that not only was he listening in on conversations that were not directed to him, he was beginning to translate some of the words we said on the keyboard. For example, one afternoon someone approached me to tell me about a fight that had occurred between Sherman and Austin. Kanzi listened, then went to his keyboard and said *Austin* and gestured *go*, pointing in the direction where Austin and Sherman were housed. Another time, someone mentioned in passing that Kanzi had learned how to turn the light on and off. He immediately went to his keyboard and hit the key for light and then gestured toward the light switch. In response, we had to do what many parents do when they don't want their children to overhear; we began to spell out some words around Kanzi. Kanzi, like most children, recognized that we were doing this to avoid his listening and simply began to listen all the harder.

At this point we decided to test Kanzi's speech comprehension carefully and with controls, just as we had previously done with Sherman and Austin. We didn't train Kanzi for this test, nor did we offer rewards for correct answers. We simply showed him an array of three photographs and lexigrams and then spoke the word to indicate which one he was to give us. We performed three testing sessions, in which our requests alternated between spoken English and lexigrams. The tests included thirty-five different items, used in 180 trials in English and 180 with lexigrams. Kanzi scored 95 percent correct on the lexigram trials and almost as well, 93 percent, on the English trials. We were able to determine that Kanzi understood 150 spoken words at the end of the seventeen-month period. By comparison, Sherman and Austin were correct 98 percent of the time they were shown a symbol and asked to find the corresponding photo from an array of two or three pictures. However, when we used spoken English to ask for a particular photo, they appeared to guess. They hesitated, tried to look for cues, and asked us to turn the keyboard on. When all else failed, they would simply select the nearest or most interesting picture. Their overall score was near 30 percent, as one would expect if they were simply selecting the photos randomly.

Kanzi's ability to comprehend spoken English is part of the explanation of the larger phenomenon, namely, his humanlike acquisition of language capacity. Comprehension aided the emergence of the productive skill in Kanzi, as it does in humans, central to which is the understanding that words and lexigrams are referential and can be used as a mode of symbolic communication.

This discovery, another first for Kanzi, clearly would force us to rethink our ideas even further about language and about human uniqueness. "If an ape can begin to comprehend spoken English without being so trained, and was able to do more than emit differential motor responses on cue, it would appear that the ape possessed speech and language abilities similar to our own," I later commented with my colleagues, in a scientific paper. "Even if the ape was unable to speak, an ability to comprehend language would be the cognitive equivalent of having acquired language."[5]

———

The amount of time and effort that had gone into the first four years of rearing and testing Kanzi had exceeded that in any other published study I had undertaken, and the resulting body of data was the most extensive. I felt ready to present what I considered to be a strong case for a reevaluation of the language capacities in apes. Five years had passed since Herbert Terrace had published his famous paper in *Science,* in 1979, which concluded that, despite many claims, no one had demonstrated language in apes. I therefore selected *Science* as the most appropriate journal to which to submit a report of the work.

I prepared a manuscript titled "Spontaneous Language Acquisition by a Pygmy Chimpanzee," and in December 1984 sent it to the editorial offices of *Science,* in Washington, D.C. I believed that the findings with Kanzi should at least reopen the door to the consideration of language skills in other species. "These results indicate that the propensity of the pygmy chimpanzee for the acquisition of primitive language skills is considerably in advance of that yet reported for other apes," we wrote. "Language acquisition in the pygmy chimpanzee seems to be accompanied (and facilitated) by the ability to understand spoken English." We pointed out that this was the first report of an ape with these abilities, and that the results had important implications for the evolution of language capacities.

Three months later, we received a rather curt rejection letter from one of the staff editors at *Science.* Two researchers had reviewed the manuscript. The first reviewer acknowledged that we were reporting important new information that was potentially significant to progress in the field. However, the review followed this with a blizzard of specific criticisms which, it suggested, must be addressed if the paper was to be published. The second review asserted that our claim for spontaneous language acquisition in a nonhuman species was not new.

How could anyone who knew anything about the field suggest that spontaneous acquisition was not novel? The reviewer stated that he/she could not envision any way in which a lexigram system could lead to the emergence of linguistic communication, thus there could be no validity to our findings, regardless of the data we presented. I wrote to the editor at *Science* and pointed out the second reviewer's apparent bias, and asked that the rejection decision be reconsidered. In mid-April we received a second rejection letter.

In the world of scientific publication, *Science* represents one of the two most widely circulated and respected general journals. The second one is *Nature,* which is published in London. We therefore recast our manuscript, to take account of some of the suggestions the first reviewer had made, and mailed it to London in May. In July we received a letter from an editor at *Nature,* saying "After careful consideration . . . we have concluded that publication in *Nature* would not be appropriate."

We had no better luck with the *Journal of Comparative Psychology,* whose reviewers displayed the same mix of general negativism and incredulity in their comments with which we were by now becoming quite familiar. Finally, I submitted a much more detailed and extensive revision to the *Journal of Experimental Psychology,* along with the previous reviews I had received, so that the editor could see the difficulties that lay ahead. Most, if not all, of the other reviews appeared to have been written by people in the field of ape language. It seemed that, for the most part, they simply refused to believe what was being said about Kanzi. They did not raise methodological or philosophical objections; they simply denied the validity of our findings outright.

The editor of the *Journal of Experimental Psychology* responded by sending the manuscript to well-recognized experimental psychologists who were not in the field of ape language. All reviewers accepted the validity of the basic phenomena and the data that documented it, and the article was published.

Terrace had done the field of ape-language research a favor with his 1979 *Science* paper. But instead of accepting what he said in it, which was that many things that had been accepted initially without question needed to be readdressed more carefully, many had inferred that apes were incapable of language capacity. Period. Few were therefore prepared to try to understand the significance of what Kanzi was doing.

Meanwhile Kanzi continued to acquire new symbols, to understand more complex sentences, and to form combinations that appeared to have some regularities to them. Knowing that these skills needed systematic study, I solicited the help of Patricia Greenfield, a linguist at the University of California, Los Angeles, who had worked with me previously on an analysis of Sherman and Austin's data.

Patricia had done pioneering work in language acquisition in children, demonstrating, against prevailing orthodoxy, the great communicative content of single-word utterances in infants. She had pointed out that although single words by themselves might convey limited information, they were usually uttered within an informative social context, including gestures of various kinds. As a result of her insights with children, Patricia became one of the few linguists to support language performance in apes, where single-word utterances constitute the great majority of communication. As with children, the single words of apes were typically accompanied by other forms of communicative information. For those prepared to listen, ape-language research derived a degree of linguistic credibility from Patricia's theories that it might not otherwise have enjoyed.

The problem I faced at this point, however, was how to establish credibility for what Kanzi had achieved. For me, the massive amount of data we had collected carried undeniable conclusions, and yet, people seemed unwilling even to consider them. "What can we do with this data that will help people understand the kinds of things that Kanzi is able to do?" I asked Patricia one day. We were poring over the pages and pages of novel combinatorial utterances that now characterized part of our database on Kanzi. "There *must* be something we can do," she replied emphatically. "Let's start classifying these combinations and see what we come up with."

— 6 —

Inside Kanzi's Mind

PATRICIA Greenfield disappeared for a few days, working nonstop to take in all the data that were available regarding the contexts surrounding Kanzi's utterances. Unlike previous studies of apes who employed symbols, we had recorded the events surrounding Kanzi's every utterance, as well as the utterance itself. It is not especially helpful simply to know that Kanzi said *Austin hamburger*—is he calling Austin a hamburger? It is helpful to know that he said this in response to being asked what he wanted to do after he pointed to the door. It is also helpful to know that after he said this, I asked, "Do you want to take some hamburger to Austin?" to which Kanzi responded with a chorus of positive *Waa* vocalizations. I told him he could get some hamburger out of the refrigerator, which he did. Then he gave it to me to carry as we went to visit Austin. When he arrived, I asked him if he still wanted to give the hamburger to Austin and he vocalized a positive *Whuh*, took the hamburger from me, and handed it to Austin. With details such as these recorded for all of Kanzi's combinations, Patricia was able to make sense of what Kanzi had been trying to say.

When she returned, she said that she felt there was evidence of syntactical structure in Kanzi's multiword utterances. I was surprised, as such structure was not apparent to me from talking with Kanzi. Patricia pointed out that you would not expect it to be, since the patterns appeared only when you looked across many utterances that were dissimilar in meaning. She was anxious to write a scientific paper with me on this point. I hesitated, as I knew that we would have a difficult time convincing people that Kanzi used syntactical structures, given the reception the work had already received.

Most linguists were not prepared to grant language to another species, for such a view would undermine the very foundation of their field. Linguists assume that the structure of the human mind will eventually be laid bare by linguistics. They believe this will happen once linguistics reaches its goal of describing the underlying structures that unify all languages and drive language acquisition in all children, regardless of their culture. If apes, too, can be said to learn language, then language cannot be uniquely human and the goal of using it to characterize the basic structure of the human mind can never be attained.

Patricia had been challenging the assumptions of linguistics with child data, and now she was ready to do so with ape data. From where she stood, linguistics as a discipline was obsessed both with syntax as the mark of language and with its Chomsky-inspired certainty that humans are unique in possessing language abilities. At the very least, this sort of close-minded thought ignored evolutionary reality: We are sibling species with chimpanzees, with whom we share 99 percent of our genetic blueprint. In principle, therefore, the likelihood that chimpanzees and humans share at least some elements of language competence is rather high.

———

By its nature, the major body of linguistic theory deals with modern language, specifically its acquisition and structure. Few linguists appear to have realized that this perspective may distort the way questions are asked about the *origin* of language and the nature of its precursors in apes. It may be inappropriate to use the model of modern language structure as a guide for finding evidence of language in its earliest stages. For instance, it should not be a surprise to state that at one point in human history, language (and its syntax) was invented, not learned. In which case, when looking for the precursors of language abilities in apes, one would be more realistic to look for the capacity for the invention of rules rather than the knowledge of existing rules. Furthermore, an evolutionary perspective of the origin of language implies that the rules would reflect the adaptation and behavior of the early humans who invented it. Similarly, if apes do have the fundamental ability to invent rules of syntax, those rules are likely to match an ape's rather than a human's mode of adaptation in some way.

This shift away from the anthropocentric, or human-centered, view pursued by strict linguists was an important conceptual guidance that Patricia brought to her analysis of Kanzi's utterances. Her experience in cultural anthropology in Africa and Central America was vital in developing this perspective. She learned at first hand how the threads of a different culture influence its entire fabric, in material and linguistic realms. Insightfully, she applied this to the assessment of Kanzi's language, insisting that we should take a bonobo's perspective, not a human's.

A second conceptual guidance we brought to our study enabled us to eschew the strict linguist's question: Do apes have language? The answer to this question, of course, is that they don't. But what would we say of a one-year-old human infant? Does she have language? A two-year-old? A five-year-old? Language emerges as the infant matures, and so it seemed reasonable for us to take a developmental approach in our study. In a chapter we contributed to a book called *"Language" and Intelligence in Monkeys and Apes: Comparative Developmental Perspectives,* we explained our perspective in this way: "We need to look for parallels between ape language and human language in the *earliest stage of development* and, having established these, see how far the apes can travel down the path toward human language."[1] In other words, we would not be looking to see if apes have language; instead, our search would be for what elements of language are present and what elements are absent at a particular stage of development.

Nevertheless, there are some fundamental criteria that a body of rules must meet if it is to be recognized as grammar, no matter what the behavioral perspective or stage of development. In the previous decade, linguists and psychologists have offered various criteria, and Patricia and I assembled five of these as central.

1. *Each component must have independent symbol status.* For example, if Kanzi uses a combination such as *Chase dog*, it must be shown that both of these words can occur in other utterances, where they each have a different meaning, such as *Chase banana* and *Dog play-yard*. If all symbols are not free to combine with all other symbols in any manner, then it cannot be assumed that their union really is a legitimate instance of combining on the part of the ape.

2. *The relationship between the symbols must be reliable and semantic.* Some combinations such as *Ice TV*, which Kanzi made in order to ask that the television be turned on and ice be brought to him while he watched it, are combinations of things that Kanzi wants, but the two symbols lack an intrinsic grammatical relationship between them. It just happened that Kanzi put them together. However, an utterance like *Matata bite*, used to comment upon the first time his mother ever disciplined him by actually biting him, illustrates Kanzi's ability to use symbols that have a meaningful relationship to each other. In such a case, unlike the *Ice TV* example, use of either *Matata* or *bite* alone would not convey half of the message. This is because the message is in the combining and the consequent relationship that is structured by the combination of one symbol with the other to convey a special meaning in a particular context.

3. *A rule must specify relations between categories of symbols across combinations, not merely a relation between individual symbols.* This means that a grammatical relationship such as action-object must be represented in the body of data by many different action terms and many different object terms (for example, *Grab Austin*, *Hide Austin*, *Bite ball*, *Chase ball*, *Hug ball*, and so on). When this occurs with a stable order between the categories of action and object, it can be legitimately inferred that Kanzi has some understanding of these categories themselves. Without such an understanding, he would be unable to order the combinations consistently by putting the action first and the object second across many different individ-

ual symbols, all combined in different ways. In looking at Nim's data, Terrace found ordering rules, but these rules were tied to specific symbols, some that Nim always put in the first position and others that he always placed in the second position. Kanzi, by contrast, formed combinations like *Grab Austin* and *Austin go*, illustrating that a symbol like *Austin* could be used in either position depending on how it functioned communicatively in the sentence.

4. *Some formal device must be used to relate symbol categories across combinations.* This means that given that Kanzi understands such broad categorical distinctions as agent and object, he must be able to demonstrate this understanding by ordering them in some reliable rule-based manner. In this case, the rule Kanzi uses is the same as that used by all other English speakers, the rule of order; place the action term before the object term. This does not mean that Kanzi is aware that he knows this rule, or even that none of his utterances ever violates it. It means only that at some primitive level Kanzi must have some sort of cognizance of the rule because he tends to use it.

5. *The rule must be productive.* This simply means that the rule must be able to be applied to new situations and to function communicatively in those situations from the first use. Thus, once Kanzi has learned the action-object rule, he should be able to apply it to produce appropriate combinations in new situations. An example of this is Kanzi asking someone to play *Tickle ball,* meaning to tickle him by rubbing the ball all over his body. This was a game that Kanzi invented, not us, as he loved the feel of the ball over his body. He followed the correct ordering rule as he made this new combination, suggesting that for him, the rule had become productive.

In a scientific paper on our work, we explained that "different subsets of criteria have been emphasized by different investigators."[2] We had deliberately set ourselves a tough goal to reach.

———

Throughout the language program with Kanzi we kept a continuous record of his use of the keyboard, his gestures, and his accompanying behavior. For the formal analysis of structure, we chose a five-month period, beginning in April 1986, when Kanzi was five and a half years of age. During that time, we recorded Kanzi's 13,691 utterances, just a little more than 10 percent of which comprised two or more elements. About half of these multiword utterances met the criterion of spontaneity (that is, not produced in response to or in partial imitation of a caregiver), and so we finished up with a corpus of 723 combinations. This is a considerably larger body of data than any child studies have used to investigate grammatical development in humans.

We had already satisfied criterion 5, that of productivity. And we knew that Kanzi's use of lexigrams met criterion 1, that is, independent symbolic status. From a very early age, Kanzi demonstrated an understanding of a one-to-one relationship between a symbol and an object or action. And we were satisfied that his use of the "go" gesture fulfilled the criterion of an independent symbol, just as such gestures do in deaf children who invent a sign language. In our study, therefore, we focused on criteria 2, 3, and 4, the reliability of the relationships between symbols used and the presence of a rule-based ordering to the symbols. We set out to examine whether there was a structure in Kanzi's utterances.

One very clear rule that emerged during the five-month period concerned combinations that involved doing something to an object, such as Kanzi hiding a peanut or biting a tomato. During the first month, Kanzi showed no particular ordering of such symbols; sometimes he put the action first, sometimes the object. *Hide peanut* occurred just as often as *peanut hide*, for instance. But thereafter he began to follow a rather strict order, that of putting the action first and the object second: *hide peanut*, *bite tomato*, and so on. Kanzi apparently learned this rule from us, as we would touch lexigrams simultaneously as we spoke, and English generally employs an action-object ordering rule. Human children learn the rule at the two-word stage, somewhere between the first and second year.

In any case, Kanzi demonstrated his ability to learn a syntactic rule from his verbal environment, and in so doing satisfied criterion 4, which requires the presence of a formal rule (for example, place action words before object words). He also met criterion 3, which required that he be able to recognize that the ordering rule operated on categories (such as action and object), rather than specific words, as the chimpanzee Nim had utilized.

More important than Kanzi's ability to learn rules, however, was his apparent capacity to invent them. Patricia identified two such rules: The first concerned the combining of gesture and a lexigram. In this case, Kanzi used a lexigram to specify an action, such as *Tickle*, then a pointing gesture was used to specify the agent. For example, if Kanzi wanted me to chase him, he first used the keyboard to say *Chase* and then pointed to me, or touched me to indicate that I was the one to chase him. Thus Kanzi's rule was if you wish to engage someone in an action, specify the action first (via a lexigram) then specify the person (via a gesture). This happened even when Kanzi was right next to me and the keyboard was on the other side of the room; if he wanted to play grab or bite, he would walk across the room to say bite, then return to gesture toward me.

Not only was this rule arbitrary, but it was very firmly in place in Kanzi's world, so much so that even being next to the intended agent did not disrupt the ordering that Kanzi had himself elected to impose upon his means of communication. It is significant that Kanzi's rule of action first (via lexigram) and agent second (via gesture) represented the opposite order of the spoken English that we used around him all the time, and therefore was strong evidence for criterion 5, that of creative productivity.

Another rule that was uniquely Kanzi's occurred in the formation of action-action utterances. The putting together of two action terms, such as tickle-bite was in fact Kanzi's idea altogether. Indeed, caretakers almost never used such combinations. In the few instances where we found that they did so, it was always in response to a previous action-action utterance made first by Kanzi. When Patricia initially looked at Kanzi's combination of two lexigrams denoting action, she saw little evidence of structure. However, the combinations were so frequent and so uniquely Kanzi's that she continued to puzzle over them. Suddenly she realized there was an order of sorts. Kanzi tended to place the actions that required a greater distance between the two parties in the first position and the action requiring closer contact between the two parties in the last position. Thus orders like *Tickle bite* and *Chase hide* were significantly more frequent than their inversions *Bite tickle* and *Hide chase*.

When apes are playing (and other primates as well), they tend to move from distal to proximal actions. For example, in a Chase hide game, Kanzi would run away a short distance and then hide. The other party was to give Kanzi a little time before chasing to the general vicinity and then finding Kanzi and tagging him. Thus, in this case, the order of Kanzi's combination reflected the order of the events of the game. This was true for other combinations like *Tickle bite* as well. When apes and other primates begin to play close together they tend to tickle first and then, as rapport is established, play-biting begins. This is somewhat like human roughhousing; it starts out gentle at first and more distal, then the contact becomes increasingly close and sometimes increasingly rough. Kanzi's action-action combinations therefore tended to follow the etiquette of bonobo play and could be said to be bonobo-centric rather than anthropocentric. Consequently, this invented rule of Kanzi's can be seen to satisfy all of the criteria for formal grammar.

Patricia's analysis produced statistically significant evidence of structure in Kanzi's utterances. He was developmentally delayed compared to a human child, but parallel syntactic structures were present. I must admit to having been surprised at the regularities that Patricia's keen perception was able to extract from the body of data that we had gathered on Kanzi.

Both rules that Kanzi invented (the "action-action" combination and the "gesture follows lexigram" combination) have to do with coordinating actions in some way. While the first reflects bonobo behavior, the second, though arbitrary, is a way of structuring two communicative channels. Why is action-based ordering important? I believe it gives an insight into the way early humans might have imposed order on language as they invented it. Language almost certainly emerged in early humans as their social life and economic subsistence activities were becoming more complex. Therefore, there would have been a premium on the ability to coordinate actions, and a simple syntax of the sort that Kanzi invented would have been an evolutionary toehold onto the more complex syntax that eventually emerged in modern humans. Moreover, the fact that Kanzi is able to invent such rules is strong evidence for the continuity theory—that is, the idea that the mind of man differs in degree from that of the ape, but not in kind. It also gives some indication of the cognitive substrate to language that might have existed in the common ancestor of humans and chimpanzees. The more we learn about what the ape brain cannot do in the realm of language, the more we can pinpoint the specific kinds of skills that must have emerged in early humans.

One of Chomsky's arguments for the innateness of language capacity in humans is the existence of common patterns within the grammars of all languages. Underlying this commonality, he suggests, is a universal grammar, the product of a unique language module in the human brain. I agree with Chomsky that *language capacity* is innate at some level. One would hardly want to attribute even partial linguistic competence to oysters, for example, regardless of their rearing. However, any argument about continuity between humans and apes adduces the fact that humans and chimpanzees share 99 percent of their genetic blueprint as a reason to expect a sharing of some elements of language capacity. Where I disagree with Chomsky is in his assertion that the structure of language is necessarily prewired in the human brain.

The analysis of Kanzi's utterances shows that the ordering of action is important in the rules he invents. This ordering is likely to have been the case in human prehistory, leading to a common syntax that is influenced by the environment. Furthermore, just as there are environmental constraints on the way tools can be structured and still function, it is plausible that there are constraints in the way language may be ordered and still be comprehensible. The fundamental similarities of human grammars may therefore be the linguistic equivalent of fundamental similarities found in the cross-cultural design of vessels constructed to carry water. When Chomsky and his followers use the term "innate," they mean innate *and* unique to humans; for me, the term "innate" is simply a mark of our biological continuity.

As I suggested earlier, it is futile to ask whether apes *have* language, as linguistic orthodoxy demands. The significance of Kanzi's possession of certain elements of language is, however, enormous. As the ape brain is just one-third the size of the human brain, we should accept the detection of no more than a few elements of language as evidence of continuity. In my view, we had done that.

———

With enthusiasm Patricia and I wrote up our results and submitted them to *Nature* in July 1987. "We demonstrate that an ape, in a communicative environment with humans, develops a productive grammar uncontaminated by imitations, and, most interestingly, invents primitive symbol-ordering rules that he has not been exposed to in his symbolic environment," we wrote in our letter of submission. In retrospect, it is clear we should not have used the word "grammar," either in our letter or in our manuscript. To the linguist it means only one thing: human grammar. We were therefore stepping into the linguists' court, and challenging the cherished notion of human uniqueness.

We had shown the manuscript to Herbert Terrace before submitting it to *Nature*, because we valued his views. He liked it, thought it important and worthy of publication, but attacked it with a volley of small criticisms. Overall, he thought we had overstated the grammatical nature of Kanzi's utterances. When the manuscript reached the reviewing process, some referees thought it too short to do justice to the work, some said it was too long for the amount of new information it conveyed, but all were either skeptical or downright scornful of our claim for grammatical structure. The manuscript was returned to us two months later, with a rejection letter.

We persuaded the editor to allow us to address some of the issues raised by the referees, and submit a new version. We should not have expended the time and effort, for the response to the new manuscript was precisely the same as previously. "Overenthusiastic overinterpretation" was the general view, with few apparently able to accept the implications of our results: that humans are not unique in possessing a capacity for language, and are therefore not as special as most would like to believe. Our second rejection letter arrived in July 1988.

A year had passed since our first submission, and we had made little progress in getting linguists to take notice of what Kanzi was able to do. Perhaps we should have used the word "protogrammar" instead of grammar, as indeed it is more appropriate. But if protogrammar is appropriate for the primitive syntactical structures that Kanzi invented and used, then so too is it appropriate for children. "Comparative developmental psycholinguistics has been plagued by a double standard," Patricia and I wrote later. "Because children ultimately develop language, their early stages are interpreted as having greater linguistic significance than the same stages in primates. When children make up novel words on a one-shot basis, it is called lexical innovation. When chimpanzees do the same thing, it is termed ambiguous."[3] The double standard, I believe, is part of the linguists' last barricade in defending the notion of human uniqueness. It is not science.

When our work was finally published, in a conference volume in the fall of 1990, it received wide coverage in newspapers and magazines. Indeed, the volume of publicity and its enthusiasm, while gratifying, was almost embarrassing. It also provided us with some response from our colleagues. "All the evidence suggests that the animals are using sophisticated ways to request things," said Terrace, damningly, to a writer for *U.S. News & World Report.*[4] "It has nothing to do with language, and nothing to do with words," scoffed Thomas Sebeok, in a newspaper article. "It has to do with communication."[5] For once Sebeok and I apparently agree: Of course it has to do with communication.

Chomsky's comment, cited in *Discover* magazine, suggested to me a lack of biological sophistication. "If an animal had a capacity as biologically sophisticated as language but somehow hadn't used it until now, it would be an evolutionary miracle," he said.[6] How do we know that bonobos are not using these abilities in complex ways in the wild, ways that are not apparent to us yet? The more we learn about them, the more sophisticated their communicative abilities appear. Would Chomsky suggest that an ape's precise control of a joystick linked to a computer game must be an illusion, because they don't do this in the wild? No, because it doesn't threaten the uniqueness of human grammar, the last bastion of the discontinuity theorists and their peculiar creationist position.

Apart from these and other solicited comments in the popular press, linguists have been oddly silent in the scientific journals. Evidently, Patricia and I had been optimistic in expecting that data would overcome prejudice.

During the time we were trying to get our syntax paper published by *Nature*, we initiated a further major test of Kanzi's language competence: namely, his comprehension of spoken English. The trial, conducted between May 1988 and February 1989, was an important milestone in psycholinguistics, because for the first time a direct comparison was made between an ape and a human child. Kanzi was a little over seven and a half years of age when he began the trial, while Alia (the daughter of Jeannine Murphy, one of my colleagues) was two years old. As the distinguished psychologist Elizabeth Bates puts it: "If we want to understand what an organism *knows* about language, isn't comprehension the best place to start?"[7] We wanted to find out what Kanzi and Alia *knew* about language at these different ages, through an exploration of their comprehension of novel sentences.

Comprehension has been relatively neglected by linguists, for two reasons. First, it is much more difficult to study than word production, particularly when tests require that infants cooperate in test situations where they may be asked to label or identify things that they are not at the moment thinking about or wishing to play with. Second, linguists' obsession with the putative innateness of language capacity has focused attention on production and away from comprehension. And yet, increasingly, psycholinguists are coming to acknowledge that comprehension is at least as fundamental as production in the acquisition of language, if not more so. It is also being recognized as immensely complex.

We usually think of language as being composed of discrete words that are assembled into phrases and sentences—this is how it appears when we hear language spoken and, particularly, when we see it written. But the sounds of language when they enter the ear are anything but discrete. Viewed on a spectrograph, a sentence looks like a continuity of sound, with intermittent jumps in amplitude, but with little clue as to where one word ends and another begins. Moreover, the same word in different sentences may look very different. For this reason, it has been unexpectedly difficult to devise a computer program that can decode human speech. And yet, before she reaches the age of one, a human child has already begun to do just that. Beginning with phrases, the child soon moves on to decoding sentences, a process that paves the way for language production.

Despite progress in breaking language down into its components—such as intentionality, rule learning, imitation, fast associative mapping, and sequencing—there is no widely accepted explanation of how language acquisition occurs. We know that children need early exposure for efficient language acquisition, but it remains a mystery how this puts into place an understanding that words are referential items and that, properly structured, they create information-rich sentences.

The Chomskian innatist view of language acquisition explains the child's gradual ability to learn to speak, despite the cacophony of language to which she is exposed, as the result of a developmental switching on of a parsing device, a brain module designed for the purpose. There is no anatomical evidence for the existence of such a structure, and the hypothesis effectively rests on a default premise: namely, that no other hypothesis offers an adequate explanation. It is true that the process of language acquisition through which most of us pass appears to be a near miracle, given the absence of what could be considered effective teaching. But that is an inadequate rationale for assuming the necessity of a unique, undetected brain structure. A plausible alternative hypothesis is that comprehension drives language acquisition.

Children, as they learn to acquire language, make some linguistic mistakes, but their number is remarkably few, and they are often on the order of incorrectly generalizing correct rules—such as "goed" for went. A minimization of production errors would occur if children came to comprehend much of what others were saying to them before they learned to speak. It has long been recognized that comprehension precedes production, both at the single-word and sentence stages. Comprehension can therefore put in place the fundamental rules for production, with context playing a large role in giving meaning to words that are heard by the child. When they are learning to talk, children do not appear to be building up a complex grammar out of single word units and an innate parsing device. Instead, they seem to be pulling apart the syntactic structure inherent in the speech around them, through the help of speakers who mark these syntactic units by their intonation and use of phrase, and by the context of what is being said. In other words, a child learns language by listening and paying attention, not by talking.

Although by no means universally accepted as the central aspect of language acquisition, comprehension is being taken more seriously than it used to be, not least in the innateness (uniqueness) debate. With Kanzi, we had an opportunity to explore the elements of comprehension that emerged as a result of his early exposure to spoken language. If his abilities were to extend beyond comprehension of single words, we would have further evidence of evolutionary continuity of language abilities.

As I described in an earlier chapter, Kanzi showed evidence that he comprehended spoken English words from a very early age. The test we did when he was four years of age showed that he could choose the correct lexigrams with sixty-five spoken words. But it was also clear to me that he understood phrases too, albeit quite simple ones. As the years passed, his comprehension appeared to expand in range and become more sophisticated.

During our day-to-day activities, we constantly attempted to explore the limits of Kanzi's comprehension. For instance, one day he and I were down by the river and, as always, Kanzi had his ball with him. Balls had been Kanzi's favorite toys since he was six months of age. He loves all sorts of balls, large and small, soft and hard, and is never really happy unless he has at least one ball with him, but preferably two or three. His favorite balls are shiny ones that look rather like the genital blossoms or attractive sexual swellings that adolescent female bonobos sport most of the time. When the other bonobos want to tease or hassle Kanzi, they will take his ball if he is not looking at it. Never has Kanzi failed to react when I say, "So and so is about to get your ball." He immediately whips around and rushes to grab his ball back. When Kanzi has five or six balls and is trying to keep track of all of them with the other bonobos around he has a real job, for one ball or the other is always rolling away where another bonobo can grab it, and as he hurries to retrieve it, others roll away from his pile. We have occasionally made videotapes for Kanzi's viewing in which an imaginary gorilla steals one of his balls and plays with it. Kanzi is riveted to the screen when such scenes appear and must rush to locations he has seen on the television immediately afterward to search for the ball. Kanzi has an uncanny memory for his balls; he can recall where he has left one days, months, and even years later.

On this particular day when we were down by the river, I decided to ask Kanzi, "Can you throw your ball in the river?" I knew this was something he had never done before and something no one would have ever asked him to do, as we generally tried to keep everything out of the water except sticks and rocks. However, I decided to violate one rule today, just to see if Kanzi could understand such an unusual request. He promptly tossed the ball in the river. On another occasion, as we were walking through the forest with his half-sister Panbanisha, I said, "Kanzi, would you please give Panbanisha an onion?" He looked around for an onion patch, pulled up a bunch, and handed them to Panbanisha. It might be thought that in such sentences all Kanzi heard was "Kanzi, xxxxx xxx xxxxx xxxx Panbanisha xx onion," and put two and two together. After all, he could not have given Panbanisha to the onion. But we noticed that, where confusion was possible, such as, "Can you throw a potato at the turtle?" he rarely made mistakes. (In this case, the mistake would be throwing the turtle at the potato.)

Most salient, however, were conditional sentences. One day, for instance, we were visiting Austin, who was busy with a task. As a reward for completing the task, Austin was given cereal, which Kanzi desperately wanted and kept requesting. I knew that Austin would become angry if Kanzi took the cereal, and told Kanzi so. While all this was going on, Kanzi was playing with a monster mask that we had brought in the backpack. Austin was very interested in the mask, so I thought I would offer a deal. I said, "Kanzi, if you give Austin your monster mask, I'll let you have some of Austin's cereal." Kanzi promptly got the mask and gave it to Austin, and then pointed again to the cereal. It had been a linguistic bargain, and Kanzi had understood.

We recorded many such examples of Kanzi's understanding complex sentences—at least, when the subject at hand was of interest to him. When Kanzi wasn't interested, he either did not understand or simply acted dumb. (I knew he was capable of being contrary when asked to do something he didn't want to do, often doing the exact opposite.) A compilation of these carefully recorded anecdotes about the extent of Kanzi's comprehension was regarded as insufficient evidence for most people, especially the skeptics. For this reason we decided we would embark on a strictly controlled study, comparing Kanzi with a human child, Alia. Our goal was not to build a complete picture of their comprehension, but rather to discover what kinds of syntactic markers, if any, they were becoming sensitive to.

———

We developed a series of strict criteria for the test, to ensure that Kanzi and Alia were not inadvertently "trained" during the trial and that we did not cue them. Very simply, in entirely separate but identical experimental settings, we planned to present them with a series of novel sentences (660 in all) and monitor their responses. Initially, the experimental setup was a little unsettling for both subjects, as the person asking the questions had to be out of sight. The questions therefore were delivered by one experimenter as a disembodied voice, from behind a one-way mirror. A second experimenter sat with Kanzi and Alia, sometimes being part of the ensuing action, but always recording what was done. This experimenter wore headphones with loud music, so he or she could not hear the sentences being directed at the subject. The sessions were also videotaped, so that they could be scored by an independent observer.

We were unsure at the outset how extensive Kanzi's and Alia's comprehension competences would be, and so we were prepared to change the complexity of what we asked as the test proceeded. It turned out that we had to increase the complexity of sentences, for both subjects. Whatever the complexity, however, we made sure that many of the sentences were unusual, such as asking them to wash hot dogs. This sometimes caused puzzlement, but mostly comprehension won through. We devised five major types of sentences (with subgroups, giving a total of thirteen) designed to test issues such as word order and complexity of structure.

By the end of the nine-month test period, both Kanzi and Alia had demonstrated a well-developed ability to comprehend all types (and subtypes) of sentences, with Kanzi scoring just a little ahead. Overall, Kanzi correctly answered 74 percent of the sentences, while Alia's figure was 65 percent. Often, when errors were made they were semantic or through inattention, not a misunderstanding of the structure of the sentence. For instance, when I asked Kanzi to "Pour the milk on the cereal," he poured the milk on the mushrooms. He performed the correct action according to the structure of the sentence, but with the wrong object. Similarly, Alia once put peaches into yogurt rather than into the tomatoes, as she had been asked.

On another occasion, Kanzi clearly mixed up words, and thought he was doing as I had asked, which was to "Put the paint in the potty." He promptly picked up some clay (a similar play object to paint), and put it in the potty. I said, "What about the paint?" Kanzi put more clay in the potty. I said, "Thank you," but "now put *the paint* in the potty." Kanzi clearly thought me a little dumb, and so brought me the potty and placed it right in front of my face so that I could see that he had done what I was so persistently asking.

One difficulty we had not foreseen, but which proved illuminating, concerned requests to retrieve an object from another location, such as, "Go to the group room and get the ball." If there was a ball where Kanzi was sitting when he heard the request, he often glanced at it, glanced toward the group room, perhaps touched the ball, and, 50 percent of the time, handed it to the experimenter. We thought of another way of asking the question, which linguistically is more complex: "Get the ball that's in the group room."

Structures such as "that's in the" are known by linguists as embedded phrases and are thought to be uniquely reflective of human thought in their recursive structure. In such sentences, one part of the sentence refers back to another. In order to understand the meaning, one has to know that one word refers to a specific word that occurred earlier in the sentence, and that the second word, in some way, changes the meaning of the first. For example, in the sentence, "Get the ball that's in the group room," the words "group room" refer back to the word "ball" and specify a particular sort of ball. Since parents have not been observed actively explaining how "recursion" works to children, yet children understand such sentences, linguists assume that an innate grammatical device permits them to decode such embedded references.

For Kanzi, the embedded phrase helped clarify the request, not confuse him. When he heard such sentences he tended to set off in the direction of the required location, with rarely a glance at a decoy object in front of him. He scored 77 percent correct answers with such requests, compared with 52 percent for Alia.

Oddly, one of Kanzi's greatest problems was with sentences that were grammatically the most simple, in which he was asked to do something with more than one object. For instance, if I said, "Give Sue the hat and the potato," he would readily give me the hat, but not the potato. He would then give me the potato if I reminded him. Sometimes Kanzi forgot the first object, other times it was the second one. It seemed that as soon as he focused on one, the other was forgotten, suggesting that the problem was one of short-term memory. There was no inherent relationship between, in this case, the hat and the potato that helped him respond correctly. If I said, "Put the hat on the potato," he had no difficulty remembering both objects, though he might simply place them side by side rather than putting the hat on the potato. Such errors revealed the way in which sentence structure aided comprehension, both for Kanzi and Alia.

Some things simply baffled Kanzi, however. For instance, he was completely at a loss with "Can you put the Coke can in the trash can?" Not only did the multiple use of "can" confuse him, but he could not come to terms with the concept of putting trash in special places. Bonobos leave trash where they make it. To put it in, say, a backpack to take back to the lab, where we would put it in another container, the trash can, was quite beyond bonobo comprehension—literally. The important thing to note about such an error is not that Kanzi could not understand the words or the structure of the sentence, but rather that he had no means of determining the set of things it was that made something deserve the label "trash." The only common element lay in the fact that they were all things we did not want. Many of them were things that Kanzi was interested in, however. Similarly, things we were often interested in, such as computer diskettes, seemed little more than trash to Kanzi, as he had no use for them.

The test demonstrated what I had believed to be correct: namely, that Kanzi's comprehension went far beyond single words and simple phrases. In fact, I realized that I had under-estimated his abilities rather than overestimated them. In a formal description of the project, my colleagues and I wrote: "These data support the view that both Kanzi and Alia were sensitive to word order as well as to the semantic and syntactic cues that signaled when to ignore word order and when to attend to it. . . . The similarity between the two subjects is all the more remarkable in that, while able to comprehend sentences, neither subject was as yet a fluent speaker."[8]

No one who sees Kanzi under these test conditions fails to be impressed, but I have found some people's reactions curious. For instance, during one presentation at a meeting of the American Association for the Advancement of Science, in which I presented a videotape of the comprehension test just described, one person asked how it was that I could suggest Kanzi understood language, when he did precisely what I asked him to do. Since the main point of the presentation had been to demonstrate that Kanzi is able to do what he is asked to, I was at a loss as to how to answer this query. It seemed that, from the questioner's point of view, comprehension had little, if anything, to do with language. From my perspective, comprehension was the essence of language, and was far more difficult to explain and to achieve than production. Comprehension demands an active intellectual process of listening to another party while trying to figure out, from a short burst of sounds, the other's meaning and intent—both of which are always imperfectly conveyed. Production, by contrast, is simple. We know what we think and what we wish to mean. Speech production is simply a matter of mechanically transforming our thoughts into speech sounds. We don't have to figure out "what it is we mean," only how to say it. By contrast, when we listen to someone else, we not only have to determine what that other person is saying, but also what he or she means by what is said, without the insider's knowledge that the speaker has.

Bonobos are more vocal than common chimpanzees, and Kanzi is no exception. In fact, when he was still quite young, I began to notice that he was exceptionally vocal, even for a bonobo. At times he even seemed to be trying to imitate some vowel sounds. For instance, when I gave him peanuts, I would say, "Kanzi, would you like peanuts?" And Kanzi would vocalize, "e-uh," a two-step sound, like "peanut," without the consonant. Similarly, with melons, he vocalized "eh-uhn," again a two-step sound, like melon. I made the same sounds back to Kanzi, to encourage him. I talked about my observations with my colleagues, documented what I heard, and tried to get them to do as I did. It proved to be a difficult exercise, partly because some people had difficulty hearing the sounds, and we all sounded rather different when we made them. In short, we could make sounds that were within Kanzi's range, but we really did not know how to construct a communication system that we could all readily understand at that time.

As I was carefully studying tapes of Kanzi and Sherman and Austin, I was impressed by the way Kanzi vocalized while communicating in other modes, such as using the keyboard and gestures. Symbols, gestures, and vocalizations seemed to be integrated as a communicative package. I decided I needed to look more carefully at the nature of his vocalizations.

With my colleague William Hopkins, I recorded Kanzi's vocalizations and compared them with those of four bonobos at the Yerkes Regional Primate Research Center, Lorel, Laura, Linda, and Bosondjo, none of whom had been language trained. Using spectrographic sound analysis (which is simply a way of transforming the wave form of sound into a visual picture), we were able to identify fourteen different sounds (strictly, groups of sounds) from the body of data we collected. Ten of them were common to all five chimps in the study, while four were unique to Kanzi. The four groups of vocalizations are "Ennn," "ii-angh," "WHAI," and "Unnn." Most bonobo sounds tend to slide from one syllable to the next, but some of Kanzi's unique sounds displayed clear shifts, as in the "ii-angh group.

My impression had therefore been correct: Not only was Kanzi vocalizing more than other chimps, but he was also making novel sounds. And from our videotapes of the sound recording sessions, we could see that the novel vocalizations were nearly always a response to a question by his human companion, a response to a comment by a human companion, or a vocal request by Kanzi. What was he saying?

Although the distinction was not always sharp, we were able to discern some pattern in the use of the sounds. For instance, the "Ennn" group, which was the most frequently uttered sound, seemed to be part of a request, often accompanying a gesture. One example on tape was when Kanzi pointed to food in a nearby cooler, but his human companion failed to notice. Kanzi repeated the gesture, this time adding an urgent "Ennn." Kanzi used the "ii-angh" group most often in response to questions, and particularly questions that ended with a two-syllable word, such as *peanut.* The "WHAI" vocalization was used most frequently after a question, such as, "Do you want to hide?" or "Do you still want to get your ball?" The last of the four, the "Unnn" group, followed a variety of human queries, such as, "I was going to put some Kool-Aid in your bowl, do you want some?"

Kanzi's novel vocal repertoire challenges the widely held belief that nonhuman primate vocalization is hardwired and cannot be significantly modified. Not only are Kanzi's sounds unique among bonobos, but they are also distinctly unlike bonobo sounds. The fact that he learned this repertoire in a language-rich environment while acquiring an extensive comprehension of spoken English prompts the speculation that he is trying to imitate human speech, or at least the inflection in such speech. As he has grown older, Kanzi has gained greater control over his vocal tract and appears to continue to attempt to imitate speechlike sounds.

While I was explaining the vocalization work to a reporter for a science magazine some while ago, the reporter said to me, "They'll never believe you." To which I replied, "Science is not about doing things people will believe. It must explore the phenomena that are out there, believable or not." Had I been guided in my work only by what was thought "believable," I would not have learned that Kanzi could acquire language spontaneously, as humans do, develop extensive comprehension, as humans do, and invent his own grammatical rules, as human ancestors once did. Who knows, maybe one day Kanzi will throw away his keyboard and say, "I'm fed up with Herbert Terrace claiming I don't have language." Personally, however, I don't think Kanzi cares at all about that. I think Kanzi would say, "I'd like to meet a good-looking female bonobo, preferably one that has learned to speak."

—

As soon as we had noticed that Kanzi, at two and a half years of age, had spontaneously acquired language skills, and continued to develop them, two obvious questions were raised. First, was Kanzi unique among bonobos? And second, were bonobos significantly more endowed linguistically than common chimps? My experience with Sherman and Austin seemed to indicate that common chimps were different. Eight years of language use around Sherman and Austin had not elicited any significant comprehension of spoken words.

We addressed the first question very early in the program, by raising Kanzi's half sister Mulika in a language-rich, social environment, exposing her to the full range of lexigrams right from the beginning. She began to use lexigrams spontaneously by the age of one year, much earlier than Kanzi had. So we knew he wasn't unique.

I formed the strong impression that bonobos were different from common chimpanzees, and stated so in several scientific papers. I should have waited until we conducted the obvious test, which was to raise a bonobo and a common chimp in the same language-rich environment. This we did with Kanzi's second half sister, Panbanisha, and a common chimp named Panzee. At first it looked as if my initial conclusion had been correct, because Panbanisha began using lexigrams within a year, while Panzee did not. By the time she was eighteen months old, however, Panzee began using symbols, and went through a learning spurt. She never fully matched Panbanisha's skills, though, either in production or comprehension.

The dual lesson we learned from the project with Kanzi, Mulika, Panbanisha, and Panzee, therefore, was that chimpanzees can acquire language skills spontaneously, through social exposure to a language-rich environment, as human children do. And, again like humans, early exposure is critical. Chimpanzees do travel down the language road given the appropriate rearing environment, but they travel more slowly than humans, and not as far. As Elizabeth Bates comments: "The Berlin Wall is down, and so is the wall that separates man from chimpanzee."[9]

We must now determine where to go from here. Can we learn to live with these higher animals that are clearly no longer unfeeling, unthinking, stimulus-bound creatures of meat and bone? Can we meet them on their own terms? Can we even understand what their terms are? How shall we forge a new ethic that takes into account not only our fellow human beings and the fragile ecosystem of the planet, but the needs and wants of all manner of other sentient beings as well? Is it possible to structure a future in which not only *Homo sapiens*, but other conscious beings inherit the earth as well? We are still the creature that plans ahead the farthest, at least in any conscious sense. What kind of world should we create with our great planning skills and our newfound knowledge of the minds of apes?

—7—

Childside

I FIRST encountered Bev* and Connie in 1981, when they were seventeen and nineteen years old, respectively. Both young women had been born with severe brain damage and were profoundly mentally retarded as a result. Despite remedial language programs, neither had learned to speak. Connie, whose motor-speech mechanism was impaired, often vocalized, but for the most part unintelligibly. Bev, who suffered cerebral palsy, was even more vocal, but was also largely unintelligible. Bev and Connie were part of a small experimental program at the Language Research Center in Atlanta, where I work with Kanzi, Sherman and Austin, and the other chimps. Before they began their visits to the center, the two institutionalized young women had been judged to have reached a level of cognitive development equivalent to that of a two-year-old. But to me, and to Mary Ann Romski, who was directing the program, there appeared to be an important difference between the two of them.

The research program was modeled after the center's Animal Model Project, and therefore involved teaching speech-impaired humans to use lexigrams as referential symbols in a communicative way, just as we had taught chimpanzees to do. As I watched Mary Ann and our joint colleague Rose Sevcik work with these subjects, I became aware of subtle signs that Bev possessed a

---

*The names of the individuals who took part in the research experiments discussed in this chapter have been changed to protect their identities.

degree of comprehension that was absent in Connie. If, for instance, Mary Ann asked Bev, "Do you want a drink?," Bev's eyes might flick toward the drinking fountain. Or, if told, "It's time to go," Bev's body might make a slight movement in response. Because of her cerebral palsy, voluntary movement was difficult for Bev, and she could not prevent herself from making involuntary limb and torso movements. Mary Ann and I talked about our separate observations of Bev's apparent comprehension, and we agreed something different was going on with her compared with Connie. That difference was soon manifested in a different rate of learning lexigrams: Bev learned quickly, while Connie was slow.

I was not directly involved in Mary Ann's research program, except as an enthusiastic spectator and occasional source of advice on how to proceed when teaching problems arose. To that point in time, my work with Sherman and Austin had served as a model and trouble-shooting device for the human program, particularly in circumventing apparent learning barriers. Every one of the difficulties that Mary Ann faced in installing language in mentally retarded individuals, I had already experienced with Sherman and Austin. In every case, Mary Ann was able to overcome the problem using techniques I had used with the chimps. The rationale of the Language Research Center—to bring together language studies in humans and chimps—was vindicated by these successes. So, too, was it vindicated by my observations of Bev and Connie. Partly from seeing the effect of comprehension on Bev's symbol acquisition skills, I became aware of the power of comprehension. I also became attuned to the subtle signs that betray comprehension. When, a few years later, Kanzi began to manifest such signs, I was sensitized to detect them. The center's "childside" and "chimpside" were informing each other in important ways.

———

The Language Research Center had been established by the Georgia State University Foundation in 1981, and was a direct descendant of an innovative initiative that Duane Rumbaugh had promoted a decade earlier. He had joined the Yerkes Regional Primate Research Center in 1969, and became associate director and chief of behavior. Yerkes was (and still is) linked with Emory University. It represented the richest resource in the country for the study of primate behavior, including scientific personnel of many different disciplines. In the winter of 1970, the confluence of two factors led to Duane's initiative. First, he became increasingly fascinated with reports from Allen and Beatrice Gardner of their language project with Washoe, and from David Premack of his project with Sarah. To Duane, the prospect of learning about the components of language acquisition through research with nonhuman primates represented a potential major advance in psycholinguistics. With an ape, it would be possible to manipulate the environment and observe the effects on language acquisition. Such manipulations were not possible with human children. The second factor was a publicly announced agenda by the National Institutes of Health (NIH) to fund research that could be applied directly to social and medical problems facing our society. (The NIH is the major funding agency for biomedical research in the United States.)

"It required just a few seconds' thought to realize that if we could learn how language develops in a chimpanzee, we would surely have a leg up in learning how to cultivate language in mentally retarded children," Duane now says, modestly. "I knew that the technology we developed would be vital, both for enabling ease of use by chimps and humans, and for making economical use of manpower."[1] Duane wanted a partly automated system, which would therefore be less demanding of instructors' time and effort than the systems used by the Gardners and Premack. Knowing he had the unique resources of Yerkes to draw upon, he decided to aim for an electronic system that would bring efficiency and objectivity to the project.

Duane recruited the help of his friend and colleague Harold Warner, chief of the Biomedical Engineering Laboratory at Yerkes, and the two of them quickly focused on the idea of a computerized system, probably with a keyboard input. A linguist was needed, and so they asked Ernst von Glaserfield, of the University of Georgia, to join them. Von Glaserfield recommended that his longtime colleague Pier Pisani, a computer expert, also be part of the team. During the winter of 1970, the four men met many times to create the outlines of a system and assemble a research grant proposal for submission to the National Institutes of Health. "We were confident that ultimately the project would lead to a better understanding of human language and some of its cognitive prerequisites," Duane, Warner, and von Glaserfield later wrote.[2] A four-year grant was awarded in the spring of 1971, and the LANA Project was born.

LANA stood for LANguage Analogue, which relates in part to the symbol system devised for the project. Lana was also the name of the first chimpanzee to work with the system. Von Glaserfield was principally responsible for developing the syntax that formed Lana's symbol strings, which he called Yerkish, in honor of Robert Yerkes, who had founded the Primate Research Center in 1924. Each symbol, or lexigram, was arbitrary and stood for a single word, including verbs, nouns, and adjectives. Each lexigram was built from combinations of one, two, three, or four geometric forms, of which there were nine in all; there was a total of several hundred potential Yerkish words. Different classes of words were denoted by the use of one of three different primary colors.

A subset of the lexigrams was displayed on a five-by-five matrix keyboard. When a key was pressed and released, the lexi-

The Lana keyboard.

gram went from being dimly lit to brightly illuminated. In addition, the lexigram was displayed on a projector above the keyboard, and a sequence of keystrokes produced a sequence of lexigrams—a "sentence"—on the projector, such as "Please Machine Give Juice." Being computerized, the system could operate day and night and did not require the presence of an instructor. Lana could interact with the machine and control aspects of her environment at will. In addition, all use of the keyboard, by chimp and instructors, was stored in the computer, bringing a measure of objectivity to the recording of communicative interactions.

Lana gradually acquired a large productive vocabulary and learned to generate a series of stock sentences, by which she typically gained food, turned music on, or obtained some other object she wanted. In addition to the stock sentences, Lana occasionally generated novel sentences, such as asking for an overripe banana by saying, *You give banana which is black.* But these were the minority of her utterances. Although Lana's behavior was impressively languagelike, we eventually realized that her abilities were limited in important ways, as I described earlier. The focus on engendering productive behavior meant that Lana's comprehension was not well developed. As I described earlier, only when our work with Sherman and Austin was well advanced did we come to appreciate fully this key distinction. Nevertheless, Lana's success with the computerized keyboard system was sufficient to encourage Duane to embark on the second aspect of the project: to discover whether non-speech communication could be taught to severely mentally retarded children who are unable to speak.

There are at least 1.25 million children in the United States who are either without speech, or whose speech is severely impaired, as a result of neurological, physical, or psychological disability. For a long time, clinicians favored persisting with speech therapy with such children, fearing that the teaching of nonspeech communication might inhibit whatever latent capacity for spoken language still existed. As a result, the development of nonspeech communication systems—namely, signs of some kind—has only recently become fully established. For instance, when the first volume of *Language Perspectives*, which is now a leading journal in the field, was published in 1974, its inclusion of a section on such systems was considered a major advance on earlier publications concerning language acquisition and intervention.

Nonspeech communication systems, or augmentative language systems, can be divided into two types: aided and unaided. The terms refer to whether or not some piece of equipment is required for communication. Gestures, mime, and manual sign languages, for instance, are unaided systems, whereas anything that uses physical symbols, such as pictographs or arbitrary lexigrams, are aided systems. The LANA system is an aided system. An important advantage that aided systems offer, particularly those like the LANA system that use visual-graphic symbols, is that the signal does not fade. In spoken language, words as symbols have to be brought to mind, and then produced. And once produced, they reside only in the listener's memory. Manual gestures rely on recall of the symbol by the producer, and are similarly transitory. A visual-graphic symbol, on the other hand, is available for recognition, not recall, by the producer, and once indicated for use may remain visible for as long as is required. Visual-graphic systems therefore place fewer cognitive demands on mentally retarded individuals, thus giving them an advantage among nonspeech systems.

The key issue facing those who wish to teach symbolic communication to speech-impaired children is the absence of a clear understanding of the nature of language acquisition and how it relates to other cognitive development. The different domains of cognitive development are by no means independent, and must interact synergistically. As Elizabeth Bates and her colleagues have shown, the best predictor of a child's skill on nonverbal tasks is the level of language comprehension obtained by the child. The mentally retarded child is therefore doubly disabled, often failing to acquire language, which would promote further mental development.

Since nonimpaired children acquire language spontaneously and with ease and speed, they offer no real clue as to how impaired children might learn. The very reasonable assumption was therefore made that children with extreme learning difficulties would have to be taught a communicative system in a very structured way, rather than be encouraged to learn as other children do.

Traditionally, language interventionists have pursued two different routes to structured teaching: One is remedial, the other developmental. The remedial model supposes that children "being taught language relatively late in their lives, because they have failed to acquire it adequately in their earlier experience, no longer possess the same collection of abilities and deficits that normal children have when they begin to acquire language."[3] This very pragmatic approach attempts to teach useful communication skills as quickly as possible, without consideration of the prelinguistic or cognitive skills the child might already possess.

The developmental model, on the other hand, is driven very much by theory, specifically, theory based on the observations of nonimpaired children. According to this approach, the impaired child, in learning communication skills, must pass through the same steps that a normal child spontaneously does. But as these steps are at best incompletely understood due to the largely hidden nature of normal language acquisition, only limited guidance can be developed for teaching strategies.

Neither the remedial model nor the developmental model is of much help to those children who completely fail to speak. Such children have often been given up as hopeless cases, but it is here that the experience of teaching language skills to chimpanzees has been especially fruitful. Earlier chapters of this book have shown how our language-teaching strategies with apes evolved through time, as our experience increased and we understood more about the components of language and how they were learned. We progressed from the insights of the LANA project, through Sherman and Austin, and on to Kanzi and the other bonobos. Advances in the work with humans tracked those with chimps every step of the way, often surprisingly so.

Duane initiated the first of his studies with humans in 1975, in conjunction with the Georgia Retardation Center. Nine institutionalized individuals took part in the study. They were aged between eleven years, eleven months and eighteen years, three months, and were all profoundly retarded. Only one of them spoke in any way, this individual producing some ten word-approximations that were unintelligible to the naive listener. Traditional speech-language treatment had failed with these individuals, and they had been judged to have insufficient prelinguistic skills to warrant participation in a language intervention program. "We stacked the cards against us by working with severely handicapped individuals," Duane now recalls. "The firm expectation by staff at the Georgia Retardation Center was that they would learn nothing at all in our program either. So, if we did make progress, no matter how small, that would have to be recognized as important."[4]

Five of the original nine people in the study continued in it for more than three years, and all of them did learn how to use symbols in a simple, communicative way. The size of the acquired vocabulary varied considerably, with a range of twenty to seventy-five symbols. With LANA-based teaching, the program achieved LANA-like abilities in the subjects, namely, the apparently competent symbol production but rudimentary comprehension. Like Lana, Washoe, Sarah, and Nim, these children learned to make their wishes known and to answer simple questions by stringing symbols together. Yet they also suffered similar deficiencies. Although they could answer questions with words, they often could not respond to commands with appropriate behavior. For example, Sam might say *Yes want cookie* if asked whether he was hungry, but if asked to go to the table and get a cookie, he often failed to respond. Nevertheless, the program had achieved its aim: to test the feasibility of using the computer-based keyboard system to teach symbol association and use to nonspeaking, mentally retarded children.

Duane believes that some of the physical characteristics of the communicative system—such as the striking shapes of lexigrams, their colors, and the fact that they lit up when selected—were important in its unexpected success. This helped gain the children's attention, which is vital but difficult when teaching mentally retarded children. Duane's insight of teaching with a computer-based keyboard lexigram system had been vindicated, and it soon became possible to pursue the concept further by applying what I had been learning with Sherman and Austin.

———

An important conclusion from my work with Sherman and Austin was that chimpanzees are capable of using symbols referentially, as humans do, but that the skill emerges only through the systematic teaching of the various components of language. These components include requesting, naming, and comprehension. The individuals in the original collaborative study with the Georgia Retardation Center had acquired languagelike skills, but they lacked a strong referential quality. The LANA project had provided the model for teaching and, inevitably, LANA-like performance was the result. The obvious step to take was to implement a teaching paradigm based on what I had learned with Sherman and Austin, with the hope of developing symbolic communication among the human students. Mary Ann and Rose initiated such a program at the Language Research Center with James Pate, beginning in 1981. The program was run in collaboration with the Developmental Learning Center of Georgia Regional Hospital.

The computer-based communication system would remain essentially the same as in the earlier study, but in this case there would be far more emphasis on comprehension. As Mary Ann and her colleagues wrote in a description of the project, "Symbolic communication is a complex phenomenon requiring an individual to comprehend the meanings of symbols, to produce symbols in appropriate situations, and to use arbitrary symbols for interindividual communication."[5] Important in the new study was a set of operationally defined components of symbolic communication. They are:

1. An arbitrary symbol that stands for, and takes the place of, a real object, event, person, action, or relationship.

2. Stored knowledge regarding the actions, objects, and relationships relating to that symbol.

3. The intentional use of symbols to convey this stored knowledge about an object, event, person, action, or relationship to another individual who has similar real-world experiences and has related them to the same symbol system.

4. The appropriate decoding of, and response to, symbols by the recipients.

The most fundamental lesson in acquiring language is learning that symbols have a function in communication. As Mary Ann and her colleagues point out, for severely retarded individuals, "this first step toward learning to communicate symbolically is perhaps the most difficult to achieve, regardless of the modality in which the instruction occurs."[6] A teaching paradigm based on my experience with Sherman and Austin was designed to surmount this difficulty, moving step by step. The first step was to teach the request function.

In unimpaired children, requesting visible objects is one of the earliest communicative actions to occur in development, with other skills emerging rapidly thereafter. Language comprehension begins as early as three to six months after birth, and by one to one and a half years, children spontaneously begin to make their needs known. In children who suffer brain impairment, both comprehension and production may fail to occur. Previous attempts to train such children to communicate have typically focused on symbol production, as there seemed to be no efficient means of measuring progress in language understanding. Additionally, most teachers have assumed that if such a child learns to use words, he or she will automatically understand them. However, a major problem with approaches that fostered symbol production was that while children learned words, they often failed to use them readily outside the training situation. Thus, though the children acquired words such as "potty" or "thank you," they nonetheless failed to tell their teachers in all circumstances when they needed to go to the potty or even to ask where it was located when placed in a new setting.

Mary Ann wondered if such generalization failures resulted from a lack of comprehension skills. Perhaps these children had learned what to say to gain the teacher's approval, much as Washoe and Lana learned what to say to win the approval of their caretakers. Could it be that these children, like apes, did not really understand that words did more than get them things? Did they realize that words were being used in all manner of ways set by others, and that to "catch on" to this they had to listen and decipher the meaning or intent behind each utterance that others made? The meaning or intent of others was often very different from one's own and had to be "figured out" anew for each occasion.

Mary Ann decided to try to determine whether it was possible to give such children an introduction to language that provided them with a thorough grounding in the myriad of different ways that symbols function—as a means of asking for things, naming things, making statements about intended actions, and understanding and cooperating with others.

Four individuals took part in the study: Bev and Connie, whom I mentioned earlier, and Ruth (aged eighteen years, eleven months) and Max (aged fourteen years, one month). Like Bev and Connie, Ruth and Max had failed to acquire speech or augmentative communication, despite extensive teaching efforts. Because of their ages (relatively old) and absence of progress in traditional language-teaching regimes, the four individuals were reckoned by their caregivers to have little chance with the Sherman and Austin regime, either. Mary Ann and Rose, however, were confident of making progress.

In order to establish as strong a motivation as possible in the teaching regime, Mary Ann and her colleagues used favorite food items as objects to be requested. This would provide a strong connection between the use of a symbol and the consequences of that use. The subjects were to be introduced to one food item and its lexigram at the beginning of the program; when the association between lexigram and food item had been learned, a second would be introduced; and so on. Mary Ann and her colleagues established a series of stages of learning each lexigram, until the subject could accurately pick it out among a field of other lexigrams. Bev learned quickly, requiring only 125 trials to reach the required accuracy for the first lexigram. Connie and Ruth were much slower, requiring 469 and 903 trials, while Max never really succeeded, even after more than 2500 trials. He was excluded from the study at that point.

The difficulty that the subjects faced with the first lexigram paled against the experience with the addition of a second. The number of trials Connie and Ruth required for the establishment of accurate requesting with this second lexigram were 2217 and 1319, respectively. Even Bev found the task more difficult, requiring 298. Thereafter, however, the subjects found the learning task much easier. For instance, Connie required 102 and 455 trials for learning the third and fourth lexigrams, respectively; for Ruth, the figures were 262 and 337; and for Bev, 93 and 42. "They learn how to learn," observed Mary Ann, "and the quality of what they learn improves."[7] One mark of this improvement was the subjects' eventual production of multilexigram utterances. As Elizabeth Bates has pointed out, in mentally normal individuals, the transition from one-word to multiword utterances is the third milestone in the acquisition of language. (The first is the emergence of communicative intentionality and the second is the appearance of reference.) Bev, Connie, and Ruth can therefore be considered to have reached that milestone.

Mary Ann and her colleagues fostered naming and comprehension skills as the subjects learned each new lexigram by following the same procedures employed with Sherman and Austin five years earlier. Bev, Connie, and Ruth were like Sherman and Austin in that naming and comprehension skills needed to be fostered by specific training. At the time, we believed that this said something fundamental about the nature of language and how it must be put in place in ape and impaired-human brains. But as we were later to discover through our experience with Kanzi, it had at least as much to do with the teaching paradigm we used.

Nevertheless, the study was important at the time, and

demonstrated that "initiating instruction at the level of communicative request is a viable beginning for the establishment of symbolic communication in persons with mental retardation who have severe oral language impairments," concluded Mary Ann, Rose, and James. "The four operational components of symbolic communication were observed concurrently after these individuals progressed into Phase 2 of the study."[8]

Not only did labeling and comprehension skills emerge as the subjects' capacity for symbolic communication emerged, but so too did other language-related behaviors, such as spontaneous initiation of communication and an improvement in comprehension of spoken words. In Bev's case, improvement in her own speech also emerged. When Bev began the program she had been unable to comprehend the English words for any of the lexigrams she eventually was able to use. But as she learned to use the lexigrams, she also comprehended the spoken word for them, doubtless because Mary Ann and her colleagues also spoke the word relating to a particular lexigram when it was indicated. This comprehension seemed to drive Bev's attempts to vocalize the lexigram words. Two words were clear, namely, "apple" and "Bev"; for the rest, she produced bisyllabic approximations where previously her utterances had typically been monosyllabic. Although Connie eventually comprehended five spoken versions of the lexigram set, she produced only a /ma/ or /ba/ vocalization when asked "What's this?" Ruth was hindered by severe hearing impairment, and neither comprehended English words nor produced word approximations.

One of the characteristics typical of severely mentally retarded individuals is that they tend to be responders, not initiators of communication. As their vocabularies expanded, each of the three subjects began increasingly to initiate communication with their instructors. Ruth was first, after she had learned four lexigrams. Her first spontaneous request was *soda*, to which she then waited for a response. Connie was the second of the three to begin spontaneous use of her lexigrams, after she had learned eleven of them. Bev began after learning twelve lexigrams, but did so much less frequently than Connie and Ruth. When Mary Ann and her colleagues saw the emergence of spontaneous communication, they decided to respond as parents do with normal children; they treated the utterances as intentional. They therefore adopted a "What you say is what you get" principle. The unfolding of these various communicative skills as the subjects increased their vocabularies was impressive, and it reveals the power of language as its components become assembled.

Another manifestation of the power of language was an unexpected change in the subjects' social demeanor. They became more sociable and positive and, instead of waiting for things to be done for them, actively took part in their world. They threw fewer temper tantrums and were able to attend to tasks much longer than previously. "They are happier," observed Rose. "All of this we think is every bit as important as the communication skills. And basically it all can be traced to the word and sentence skills they acquired."[9] Stephen L. Watson, director of the Developmental Learning Center of Georgia Regional Hospital, said of Bev, Connie, and Ruth's transformation: "This is not typical of institutionalized people, who are complacent because they grow accustomed to having their needs met before they ask."

Rose's remark in attributing Bev, Connie, and Ruth's increased engagement with their worlds to the power of language seemed plausible, given that *Homo sapiens* is a creature of language whose social fabric is very much the product of words. But her conclusion went beyond instinct, and included a careful study. The study addressed the possibility that the subject's change in overall social demeanor might have been the result of the novel interaction to which they were exposed during the program, rather than the acquisition of language skills. Rose, Mary Ann, and Adele A. Abrahamsen compared various aspects of Bev, Connie, and Ruth's behavior with those of individuals who had been exposed to daily contact with instructors but had not been taught to communicate symbolically.

The individuals in the control group did show an increase in their sociability, as Bev, Connie, and Ruth had, but their intentional communication, rate of attentional shift, and attention complexity did not change. Significantly, neither did that of

Max, who was exposed to symbols, but failed to learn. The improvement in Bev, Connie, and Ruth in these three measures can therefore be seen as a direct result of having acquired a degree of symbolic communication. "The acquisition of lexigrams occurred as part of a larger package of developmental changes in this study," noted Rose and her colleagues, "which is suggestive that the process of acquiring the new symbols is embedded in a causal structure of changes involving several [cognitive] domains."[10]

This study presents a particularly clear insight into the connectedness of cognitive domains, and the power of language in humans in raising general intellectual achievement. I see a similar pattern of transformed behavior in language-trained apes, who appear to be more reflective, attentive, and sensitive to communication with humans. This is strong evidence, I believe, for the evolutionary continuity of the mental substrate between apes and humans.

By the early 1980s, Duane's vision of a decade earlier had already paid off. He, and later Mary Ann and her colleagues, had made entirely unexpected progress in teaching augmentative communication skills to severely mentally retarded children, by applying theoretical and practical knowledge gained from our ape-language studies. Individuals for whom all traditional methods of speech and other language training had failed, had learned to communicate for the first time in their lives, using Duane's computer-based keyboard lexigram system. "The results with the Georgia Retardation Center and Developmental Learning Center projects had exceeded all expectations," Duane now admits. "Quite simply, most people expected that the children would learn nothing at all, but they ended up learning a lot."[11] Not only had individuals in the program learned productive competence of a vocabulary of lexigrams, but most of them also generalized that competence to labeling and comprehension, and even to spontaneous communication. Most language intervention programs cannot make such claims.

Despite the surprising efficacy of the Sherman and Austin teaching approach, it did conform to what was believed a necessary aspect of language learning in severely mentally retarded individuals: namely, that language skills had to be drilled into place through some kind of structured teaching. True, since the mid- to late 1970s, language intervention practitioners increasingly emphasized a more naturalistic setting for such instruction, but the acquisition of language itself was not naturalistic. Mary Ann and her colleagues worked thoughtfully to produce a naturalistic teaching environment at the Language Research Center, including lexigram games. Nevertheless, they considered themselves active teachers, putting requesting skills in place and watching as other skills emerged.

When, in the spring of 1983, we began to realize that Kanzi had spontaneously acquired a small vocabulary of lexigrams and comprehension of a few spoken words, we were forced to rethink our hypothesis about the acquisition of symbolic communication in apes. Equally, rethinking would be necessary on the "childside." In a description of the efficacy of the instructional approach with mentally retarded children, Mary Ann and I had said, "These subjects needed to have language learning broken down into small units, just as Sherman and Austin had."[12] True, these subjects had learned symbolic communication through having language broken down into small units, but perhaps they too could acquire these skills spontaneously, without teaching, as Kanzi had?

"The observations with Kanzi had an immediate effect on our thinking," recalls Mary Ann. "I was anxious to be able to apply to mentally retarded individuals what Kanzi was telling us, and I was particularly keen to do it with younger children."[13] Most of the individuals in the Georgia Retardation Center and the Developmental Learning Center studies had been adolescents or young adults. Everything that linguists have learned about language acquisition in mentally normal individuals indicates the importance of early exposure—the existence of a critical period. Those who try to learn a second language after childhood are aware of the struggle it can be, compared with language acquisition in childhood. Mentally retarded individuals who are trying to learn a communicative system after childhood are therefore doubly disadvantaged, by age and by their cognitive impairment.

In the spring of 1983, a serendipitous event presented an opportunity for beginning to apply a Kanzi-like paradigm in mentally retarded school children. A group of local schoolteachers who were completing a graduate course at Georgia State University paid a visit to the Language Research Center, to observe the program with Bev, Connie, and Ruth. They were enthusiastic about what they saw, and intrigued by the description of Kanzi and his spontaneous acquisition of symbol use and comprehension. They said: "Have you thought about applying some of this out of the laboratory, to children in school, for instance?" There is often a suspicion of experimental programs in schools, not least because researchers all too frequently treat the exercise as a way of collecting data, and don't have the long-term interests of the schools at heart. Making an approach to engage in some kind of experimental program therefore has to be done with great diplomacy, but it was something Mary Ann and her colleagues planned to do at some point. The invitation by the teachers who visited that day, therefore, was more than welcome. "It was music to our ears," recalls Rose.[14]

Through further discussions with the teachers and their principals, and a scramble to find funds from various sources, a summer program was set up at the Language Research Center for a small number of students from the Clayton County Public School system. The school system benefited because it could offer a summer program for which it otherwise had no funding, and the Language Research Center benefited, because it was taking its first significant step from research in the laboratory to application in schools. The successful summer program revealed that it was possible to make significant progress with children at this severe level of retardation in an unstructured setting that zeroed in on comprehension rather than production.

After this pilot program, Mary Ann wrote a grant proposal that fully incorporated the Kanzi paradigm along with the idea of studying young children in Clayton County schools. As typically happens in such cases, a grant review committee came for a site visit, to see what the Language Research Center offered, and to hear the arguments in favor of the proposed work. This was in 1984.

"It was early in the morning," remembers Rose. "We were in the group room, sitting around a table with the review committee. One person on the committee, a very prominent individual in mental retardation research, was saying very kind things about our work, about how innovative it was, and how important for potentially changing the lives of mentally retarded individuals. He was being just *too* kind, and you knew a giant BUT was coming down the pike. Sure enough, at the end he said, 'But, there is nothing in my experience nor in the mental retardation literature that would indicate such an approach has any chance of working.' We spent the rest of the day telling him and his colleagues about the success with the chimps, and about the history of how such successes had always translated to mentally retarded individuals."[15]

Rose and Mary Ann must have made a persuasive case, because the grant was awarded.

———

The new study began in 1985, with thirteen mentally retarded male students having a mean age of twelve years, four months. None of the children had more than ten intelligible spoken words, and all had failed to acquire communicative skills by other teaching methods. Mary Ann's grant was for a two-year study, which not only would explore the efficacy of truly naturalistic learning but would also compare the effects of school and home environments. For the first year, therefore, half the children were exposed to lexigram use at home, and half at school. During the second year, they used the system both at home and at school.

Mary Ann called the program the System for Acquiring Language (SAL), and it had five components.

The first was the computerized keyboard, which by this time was smaller and more powerful than previous versions, and now included a speech synthesizer. Portability was a crucial factor, because if the system were to be a feasible part of an individual's life, it must not be cumbersome and obtrusive. Initially, the children had a Words+ Personal Voice II System, via a Unicorn touch-sensitive expanded keyboard, attached to a Votrax word synthesizer. Later a SuperWolf was substituted for the Personal Voice II. Technology has been central to the language studies from the beginning, first at Yerkes and then at the Language Research Center, and the program has benefited from the microcomputer revolution that has occurred in the past decade. Compared with what is available these days in terms of sophistication and computing power, the SuperWolf is rather primitive. But it is effective, robust, and inexpensive. No doubt great improvements are possible.

The second component was the vocabulary of lexigrams, the assembly of words that were selected in the beginning. Because teachers and parents would be interacting with the children, the written English equivalent was printed on each of the keys, to facilitate two-way communication.

Third, unlike the previous program, in which individuals were taught request skills, *no direct teaching* was to take place. The children were encouraged but not required to use the keyboard in the communication opportunities that occurred in their daily events.

The fourth component involved the children's adult partners and their active role in communication. After instruction in use of the system, the adults employed the keyboard in ways we had done with Kanzi. That is, they discussed what was going to happen, made comments, or asked questions, all using normal spoken sentences, but hitting the keyboard as well at appropriate times. For instance, a parent might say, "Johnny, let's go *outside* and ride your *bike*," where "outside" and "bike" appear as lexigrams on the keyboard.

Last, Mary Ann and her colleagues needed a way to monitor the children's progress; they achieved this through a Teacher/Parent questionnaire.

At the end of the two-year program, Mary Ann and her colleagues had amassed more than 31,000 communicative events, collected as audiotaped interactions, which were then transcribed. All the children eventually acquired a vocabulary of symbols, though the size range was large. The lexigrams in the children's vocabularies were those they had effectively chosen to learn, not those their teachers had wished them to learn, just as Kanzi had extracted from his language environment those words that were most salient to his life. All the children learned to use the symbols in clear communicative ways, often accompanied by naturalistic gestures. "In general, the SAL permitted the youths to convey specific information that their partners could respond to, thus promoting the initiation as well as the continuation of conversations and the addition of new information," remarked Mary Ann and Rose, in a report of the program.[16]

As had been observed in an earlier study of mentally retarded individuals using the Sherman and Austin instructional regime, there emerged two classes of learning patterns in this program, beginning and advanced. The four youths who displayed the beginning learning pattern acquired production and comprehension skills only slowly, and had a small vocabulary at the end, between twenty and thirty lexigrams. The advanced learners acquired large vocabularies (some with more than two hundred lexigrams) and developed comprehension and production simultaneously and rapidly. Individuals in this group also learned to use combinations of lexigrams and other symbolic skills, including the recognition of printed English words and categorization.

Mary Ann and Rose were able to identify the underlying cause of the difference in performance between these two groups by comparing results of certain language and cognitive tests that had been administered prior to the study. "The salient factor that distinguished the two groups was the speech comprehension skills they demonstrated at the onset of the study," they noted.

We suggest, then, that individuals who comprehended speech prior to the onset of the study readily extracted the critical visual information from the environment, paired it with their spoken language knowledge, processed it, and produced symbolic communications. Individuals with limited comprehension abilities apparently were confronted with a different task. They had to segment the visual component of the signal, develop a set of visually based symbol experiences, process the visual information, and then first comprehend and later produce symbolic communications.[17]

The importance of comprehension in the process of language acquisition has emerged repeatedly, both in ape and in human studies.

Early in the Clayton County Public Schools program, teachers and parents asked Rose to include in the lexigram vocabulary words that help mediate normal social interaction, such as *please, thank you, I'm finished, help, yes, no,* and *goodbye*. The children incorporated these words into their productive vocabulary surprisingly quickly. These social-regulative words were used frequently in the context of social interaction, but there was no overall increase in the use of the SAL as a result. People who interacted with the children very much appreciated their ability to communicate in a way that had the semblance of being "more like a normal child." This is important, because one of the aims of giving mentally retarded children the ability to communicate is that they should be able to become more a part of the outside world. Strangers' perceptions of such children are significant in the children's acceptance in that outside world.

Just as Bev, Connie, and Ruth had become more sociable through their exposure to language training, so too did the children in the Clayton County study. Rose recalls that the mentally retarded children in the study at one school, although they were mainstreamed for lunch times, regularly used to segregate themselves as a group, eating together at one table. "Within a few months of the beginning of the study, our kids were everywhere," recounts Rose. "They were talking to their regular education peers, ordering food in line like everybody else. What had happened? They now had a way of communication among themselves and with others."[18]

One day she observed Bob, who was part of the study, talking to his peer tutor, Ralph, using the SAL. A friend of Ralph's came up and started talking to Ralph while ignoring Bob. Ralph said that if his friend wanted to join in the conversation, he would have to use the SAL. Hesitantly, the friend began to do so, and soon said to Bob, "Let's go *outside* to the playground." Bob said yes and the three boys went out as a normal social trio, to play. "These findings highlight the role the SAL may play in mediating or advancing interactive skills between peers," comment Mary Ann and Rose. "Speech output communication devices may be one important means of enhancing social interactions with nondisabled as well as disabled peer communication partners."[19]

Stories such as those of Bob and his friend are common among the Clayton County children in the study, and give substance to the once unthinkable idea that such children can indeed function in society, and can even contribute in the workplace. To this end, Mary Ann and her colleagues have established Project FACTT (Facilitating Augmentative Communication Through Technology), a collaborative effort between Georgia State University and Clayton County Public Schools. Soon, children who not long ago would have been considered as beyond hope of help, will be doing useful jobs and communicating with fellow employees and their employer. That represents enormous progress from the idea that took Duane "just a few seconds of thought" to come up with in the winter of 1970. "Our understanding of language through the study of chimps has overshot our expectations by a thousand percent," Duane now says. "I had no idea that chimps would one day be shown to acquire language spontaneously, as human children do, nor that the ape work would inform so powerfully what can be achieved with mentally retarded children."[20]

There is still a great deal to be learned about how language interacts with the cognitive development of mentally retarded children, but all the evidence points to the importance of engendering language skills in them. And if Mary Ann can achieve her goal of getting ever younger children—even toddlers—into SAL programs, further progress is surely inevitable. The mother of one of the Clayton County children once said to her: "You've done so much for my son, but, oh how I wish you'd had him when he was two. Who knows how much you could have done for him by now?"

— 8 —

Pan, the Tool-Maker

A T the Wenner-Gren conference we were finding that one by one the great "Definitions of Man" were falling. The belief that man alone can transmit cultural acts had vanished with the early observation of potato washing in Japanese macaques by Junichiro Itani; once this custom originated, mothers passed it on to their offspring. The belief that man alone can make tools had gone by the wayside in the first years of Jane Goodall's observations in the field. The belief that man alone has language has taken a little longer to fall and is more controversial, but my fellow scientists at "the castle" were beginning to recognize that this, too, was to become a controversy of the past. Nick Toth, the resident expert on stone-tool construction, was wondering if another monolith might fall. Are humans really the only creatures that can use tools to make other tools, he wondered?

When Nick initially posed this question to me, I was uncertain as to whether he really wanted to find out, or whether he was trying, like so many before him, to shore up yet another concise description of man that could neatly distinguish humans from other creatures in a completely definitive manner. It was the end of a long day at the conference, and we had all eaten dinner at a restaurant in the nearby town. A good measure of wine had been drunk, light-hearted banter had mixed with shop talk, and some members of the group had entertained us with a bout of spontaneous music making. As well as being a fine archeologist, Nick turned out to be a gifted and enthusiastic musician. On our return to the conference hotel, I was sitting near the rear of the bus and Nick was in the very back, legs stretched out, arms folded across his chest, eyes closed, apparently asleep. Suddenly, he opened an eye and beckoned me to join him. "I have something I want to ask you," he said. "Do you think Kanzi could learn to make stone tools, the way early humans did?"

I hadn't talked much to Nick by that point in the conference, and I had formed the impression that he was an archeologist who didn't like psychology. His question seemed to come out of the blue, and was something I had never thought of trying. From my long experience with chimpanzees, I had gained a great respect for their abilities. I knew that common chimps and bonobos were skilled manipulators of objects in experimental situations, and of course was aware of the many kinds of tool behavior that common chimps display in the wild. My already keen respect for their abilities was increased by the conference presentations of Bill McGrew and Christophe Boesch. Bill pointed out that not only do chimpanzees make tools, but they essentially have a tool kit, meaning that they use similar tools for different ends (for example, they use sticks for termite fishing and as weapons) and they accomplish similar ends with different tools (for example, one chimpanzee was observed to use four different tools in the process of extracting honey from a bees' nest). Christophe described the complexity of tool-use for nut cracking in the Ivory Coast chimpanzees. These chimps use both a hammer and an anvil, often for as long as two hours per day. Hammers can be either suitable wooden clubs or rounded stones. The anvils are generally indentations on stones or tree limbs. Both hammers and nuts must be carried to these anvils, which themselves are part of a surface that cannot be transported. The tools and nuts are regularly carried over a hundred meters to anvils. The wooden hammers are constructed by the apes themselves and the stone hammers are carefully selected and in high demand. Christophe has even observed instances in which mothers appeared to be teaching their infants nut-cracking techniques.

But making stone tools is quite another matter. It seemed light years beyond what apes were currently doing. No one had ever seen a chimp intentionally make a flake. Some of Christophe's chimps had accidentally broken off stone fragments while attempting to crack nuts, but no chimps had ever been seen to use a serendipitously produced flake as a tool. Indeed, even I could not make a worthwhile stone tool, and I had had Nick there to teach me.

"That sounds like a fascinating idea," I replied to Nick. "But isn't stone tool-making a bit advanced for apes?" I had never seen Kanzi, or even the more dexterous Sherman and Austin, attempt to make stone flakes. And, like most people, I assumed stone tool-making was a *human* activity requiring *human* skills. Nevertheless, I was intrigued. "What do you have in mind?" I asked. Nick sat up, and quickly explained.

In 1949, the British anthropologist Kenneth P. Oakley published a classic book, *Man the Tool-Maker.* This short volume encapsulated what was widely held to set humans apart as unique: "Possession of a great capacity for conceptual thought, in contrast to the mainly perceptual thinking of apes and other primates, is now generally regarded by comparative psychologists as distinctive of man," he wrote. "The systematic making of tools of varied types required not only for immediate use but for future use, implies a marked capacity for conceptual thought."[1] The notion of man the tool-maker struck a receptive chord in the minds of anthropologists, and in society in general: Alone among the world's species, tool-making *Homo sapiens* fashions an elaborate culture and manufactures a powerful technology, through which the world has been forever changed.

When Jane Goodall in the early 1960s reported her observations of chimpanzees making simple tools for harvesting ants and termites, human uniqueness appeared to be threatened. In their tool-making, chimps broke twigs from trees, stripped off the leaves, and then inserted the "fishing stick" into termite mounts or onto ant trails. Tool-using of various kinds has been observed in many creatures, from sea otters, to sand digger wasps, to Galápagos finches. But Goodall's chimps were doing more than *using* tools; they were *making* them. So perhaps

Homo sapiens isn't unique, after all? The anthropologists' response was, Ah, but humans use tools to make tools, while chimps just use their hands and teeth. Human uniqueness was thereby preserved.

The ability to make and use tools has long been considered part of the evolutionary package that transformed an apelike creature into the human species. Charles Darwin argued that stone tools were vital as weapons and provided a substitute for the teeth and claws possessed by other large predators. "The free use of the arms and hands, partly the cause and partly the result of man's erect position, appears to have led in an indirect way to other modifications of structure," he wrote in his 1871 book, *The Descent of Man*. "The early male progenitors of man were . . . probably furnished with great canine teeth; but as they gradually acquired the habit of using stones, clubs, or other weapons, for fighting with their enemies, they would have used their jaws and teeth less and less."[2]

A century later, when the human prehistoric record was much richer than in Darwin's time, anthropologists continued to view tools as an integral part of the evolutionary wedge that separated humans from apes, specifically in arming humans as hunters. "In a very real sense our intellect, interests, emotions, and basic social life—all are evolutionary products of the success of the hunting adaptation," said two prominent anthropologists at a landmark scientific meeting in 1966. "Human hunting is made possible by tools."[3]

The shift to becoming a tool-maker—and specifically, a maker of stone tools—has been seen as central to what differentiated humans from apes in an evolutionary sense. By definition, therefore, the very first members of the human family must have been tool-makers. This assumption has been challenged in the past several decades by the many discoveries of ancient human fossils and artifacts from East Africa, and by molecular biological information on modern humans and apes. The first members of the human family are now known to have evolved at least five million years ago, perhaps as many as eight million. And yet, the first recognizable stone artifacts date only to two and a half million years ago. The arrival of these stone tools in the record coincides with the first appearance of the genus *Homo,* which eventually gave rise to modern humans. The obvious assumption is that this new, large-brained member of the human family was the tool-maker; and that tool-making and increased brain size were linked in some way.

Whatever the truth is of that assumption—and it is difficult to imagine how it might firmly be substantiated—it raises an important question for archeologists and psychologists about the earliest tool-makers: Were they doing something that was beyond the cognitive capability of apes? Or were they merely bipedal apes who were applying their apelike cognitive skills to non-apelike activities?

Nick told me that he had been musing over this question for a long time, and had once approached the Gorilla Foundation with a proposal for an experiment, to see if gorillas could learn to make stone tools. The foundation was busy with other projects, and turned down Nick. It was, he said, an idea in search of a collaborator. His proposal was to motivate Kanzi to make stone flakes, not teach him with structured lessons. "We want to avoid the criticism of classical conditioning," he said. He suggested we would need a box with a transparent lid. Something enticing would be put in the box, and the lid would be secured with a length of string. Kanzi could be shown by example how to make flakes—by knocking two rocks together, for which the proper archeological term is hard-hammer percussion—but there would be no active teaching, no shaping of his hands, no breaking the task down into component parts.

A little more than a decade ago Richard Wright, of the Bristol Zoo in England, had taught an orangutan to make stone flakes and to use them to cut string to reach a food reward. But Wright had taught the orangutan each component of the task separately. He had also arranged it so that one of the rocks was secured to a plank of wood, not held in the hand as tool-makers did, and Kanzi would. Although Wright's experiment stands as an important part of primatological research, what Nick had in mind would go further. Wright had taught the orangutan the behavioral components of the task, while our approach would be to impart the conceptual components.

I made some suggestions to Nick about how the design of the food box, or tool site as we came to call it, could be improved. Nick had underestimated Kanzi's inclination and ability to tear flimsy objects apart, especially if there is food inside. We talked a little about how to carry out tool-making activities so that Kanzi might want to emulate us. And Nick promised to get in touch with me after we returned to the United States. This he did within a couple of weeks, and I told him that we had made a tool site to his specifications. A week later he arrived at the Language Research Center with fellow archeologist Kathy Schick, their truck laden with a thousand pounds of rock.

———

In 1989, Tom Wynn, an archeologist, and Bill McGrew, a primatologist, published a paper called "An Ape's View of the Oldowan." The word Oldowan is the name applied to the earliest known stone-tool assemblages, which were found in Africa and date back to two and a half million years ago. The question Tom and Bill addressed in their paper—"When in human evolution did our ancestors cease behaving like apes?"—was essentially the same as Nick's.[4] In other words, given the right circumstances, could apes make Oldowan tools? Tom and Bill's approach to the question was not experimental, as Nick planned, but instead involved examining the skills required to make Oldowan tools, and then seeking signs of such skills in aspects of chimpanzees' lives.

The artifacts that make up Oldowan assemblages were produced from small cobbles, and they include about half a dozen forms of so-called core tools, such as hammerstones, choppers, and scrapers, and small, sharp flakes. The tool-makers were assumed to have had mental templates of these various tool types, which they used to produce tools for a range of different functions. The tools are often found in association with broken animal bones, which sometimes show signs of butchery. The clear inference is that, beginning about two and a half million years ago, our human ancestors began exploiting their environment in a non-apelike way, by using stone tools as a means of including significant amounts of meat in their diet.

Until quite recently, archeologists argued that the earliest tool-makers lived lives analogous to those of contemporary hunter-gatherers. In other words, they organized themselves into small, mobile bands, established temporary home bases, and divided the labor of hunting and gathering between male and female members of the band. This was a very humanlike way of life, albeit in primitive form, and most definitely unlike that of an ape.

In recent years, however, a reexamination of the archeological evidence has changed this picture dramatically, making it much less humanlike and more apelike. There is no question that these early members of the human family made tools and butchered animal carcasses. But there is considerable debate over the extent to which they were active hunters as opposed to opportunistic scavengers. And the notion of home bases and a division of labor between the sexes has been abandoned as untenable. The earliest tool-makers are now viewed as bipedal apes who lived and foraged in social groups in a woodland/savannah environment, as baboons and chimpanzees do, and whose repeated use of specific locales for butchery activities created the archeological sites that are being uncovered today.

An equally important shift of perspective has taken place regarding the tool assemblages themselves. In her pioneering work in East Africa in the 1940s, 1950s, and 1960s, Mary Leakey identified what she assumed were intentionally produced tool types. But when Nick Toth began a program of experimental archeology in the 1970s, in which he became a proficient maker of Oldowan artifacts himself, he came to a very different view. "My experimental findings suggest that far too much emphasis has been put on cores at the expense of flakes," he wrote. "It seems possible that the traditional relationship might be reversed: the flakes may have been the primary tools and the cores often (although not always) simply the by-product of manufacture. . . . Thus the shape of many early cores may have been incidental to the process of manufacture and therefore indicative of neither the maker's purpose nor the artifact's function."[5]

Nick's reassessment of the Oldowan artifacts revolutionized African archeology, and further changed the perception of the humanness of the earliest tool-makers. According to this new theory, the half dozen different tool types in Oldowan assemblages were not the product of mental templates in the minds of sophisticated tool-makers; instead, they were the results largely of the shape of the cobble from which the flakes were struck, and the nature of the raw material. The only skill required by the earliest tool-makers, therefore, was that of striking flakes off a core using a hammerstone. How cognitively demanding is such a task?

"One of the more direct ways to assess the cognitive ability employed in tool-use is through examination of spatial concepts," wrote Tom and Bill in their 1989 paper. "The spatial concepts required for Oldowan tools are primitive. The maker need not have paid any attention to the overall shape of the tool; instead, his focus appears to have been exclusively on the configuration of edges."[6] Tom and Bill identified three spatial concepts in relation to tool-making: proximity, boundary, and order. Simplest of the three is proximity, which refers to the ability to land hammer blows repeatedly in the same or more or less the same position on the core cobble. A more complex notion is that of boundary in space, which has to do with the division of a spatial field into different realms. This refers to the process of striking flakes off both sides of a cobble, yielding a bifacial chopper. The most sophisticated spatial notion is that of order, which concerns control over the location of a sequence of hammer blows.

Many Oldowan tools have just two or three flakes removed, while few have more than a dozen. Once the first flake is removed, the direction of subsequent flaking appears to be fixed. The complexity of the tools, in relation to spatial concepts of their manufacture, is therefore rather limited. What of the spatial concepts involved in the natural tool-making of chimps?

Most of the tools chimps make are of raw material other than stone, and so a direct comparison is impossible. And many of the tools' characteristics are influenced by the nature of the material. For instance, the position of leaves to be stripped from a twig is a property of the plant, not a decision of the ape.

However, suggest Tom and Bill, it is possible to find examples of the three spatial concepts in several realms of chimp behavior. For instance, chimpanzee "artists" do not scrawl randomly on blank pieces of paper. And if they are given a sheet of paper on which a geometric shape, such as a square, has been drawn, chimps tend to mark on or near the shape, thus displaying the concept of proximity. They exhibit the property of boundary in the preparation of sedge stems as termite fishing-probes. The stem is triangular in cross section, and the chimps carefully remove one of the ridges by longitudinal stripping. Finally, the concept of order is to be found in chimpanzees' nest building, a skill that is learned rather than innate, as it is in birds. When a chimp builds a nest, the choice of the first major branch to be bent provides the foundation of the subsequent pattern of inter-weaving of smaller branches and twigs.

For Tom and Bill, the conclusion was clear: "All the spatial concepts for Oldowan tools can be found in the minds of apes. Indeed, the spatial competence described above is probably true of all great apes and does not make Oldowan tool-makers unique."[7] And they respond to the positing of the "using a tool to make a tool" as a Rubicon, by pointing out that Richard Wright's orangutan learned to do just that, so the skill is within apes' cognitive realm. Moreover, they make the reasonable point that chimps have sharp teeth, and use them to fashion wooden tools; they don't need stone flakes for the job. "We cannot fault apes for not employing unnecessary techniques in making tools needed in their subsistence," they add.[8]

When they examined ape and early human subsistence strategies, including tool-use, Tom and Bill again see no great cognitive divide. For instance, anthropologists often point to two aspects of early human strategies as evidence for advanced mental powers. One is the fact that several different types of raw material may be used for tool manufacture at one site. And the second is that raw material for tool-making is often transported from as far as ten miles distant. But, say Tom and Bill, the chim-panzees of the Taï Forest use granite rocks to crack the hard Panda nuts while using softer rock or even a wooden hammer for Coula nuts, which are easy to crack. Moreover, the chimps often carry their rock hammers a third of a mile to the nut-cracking stations. This latter point is therefore a quantitative not a qualitative difference, and surely has to do with a bipedal animal's greater ease of carrying things.

Modern chimpanzees are known to be occasional hunters, whereas early humans are inferred to have hunted, although some anthropologists suggest they were exclusively scavengers. Early humans used stone tools to gain access to meat on a carcass, whereas chimpanzees use their teeth to the same effect. It is true, however, that early humans sometimes processed the carcasses of large animals, whereas chimpanzees' meat-eating is exclusively of small- to medium-sized creatures, such as small antelope and monkeys. The use of sharp cutting tools allows for butchery of large carcasses, an activity for which canine teeth are not suited.

Judging from the evidence of fossilized animal bones on archeological sites, early humans frequently broke open bones to gain access to the marrow. Archeologists characterize this as an "extractive" process, a term that can also be applied to the removal of meat from a carcass. This does not distinguish early humans from apes. For instance, probing for termites, breaking open nuts, and using leaf sponges to soak up fluids or brain material are all extractive processes, and each is part of the subsistence repertoire of chimpanzees.

"What we know about Oldowan foraging seems to be within the capabilities of apes," conclude Tom and Bill. "The details differ, but this might be expected given local differences in habitat. Oldowan foraging appears to have been that of a hominoid who lived in semi-arid, open grassland and who combined scavenging of carcasses and hunting small game. There is no evidence for a dramatic reordering of general hominoid foraging, nor for an evolutionary leap in the cognitive capabilities underlying it."[9] No one who spends many hours in close contact with chimpanzees, as I have, can fail to be impressed with their keen cognitive abilities. Tom and Bill's assessment, though regarded by some as an extreme statement of ape abilities, is surely plausible. Still, when we began the tool-making project with Kanzi, I wondered whether we were embarking on a task that required more patience and motor control than he possessed. He might understand why he needed a tool and even how to make one. Indeed, I thought this would be the easy part for him. But I was not at all certain that he could actually do so since stone tool construction requires a great deal of bimanual eye-hand motor coordination, as well as proper timing and orientation. It seemed to me that the mechanical demands of flake construction went far beyond those required for termite fishing or Panda nut cracking.

—

Nick Toth was to be Kanzi's model, as I was too incompetent a flaker to teach Kanzi anything properly. But Kanzi considered Nick a stranger, and, as an adult male, felt it his duty to frighten away all "outsiders." Thus it was not surprising that he was initially aggressive toward Nick: He puffed himself up in a great display, repeatedly rushed at Nick as he stood safely on the other side of the caged enclosure, and threw handfuls of cedar chips. Although Kanzi is gentle and even flirtatious with females, he is often aggressive to males at first encounter. I told Kanzi that Nick was there to help us, and that he should stop being aggressive. When Kanzi began to see for himself that Nick was indeed being helpful, he did calm down, but still did not fully trust Nick. When Nick got too close to the wire that separated them, Kanzi would stop watching how he made tools and begin to concentrate instead on opportunities for grabbing his shirt.

At first, we had set up the tool site outside Kanzi's cage, so that Nick could show Kanzi how it was possible to gain access to the baited box. He struck a cobble with a hammerstone, selected a sharp flake, and then cut the string securing the lid of the box. Kanzi got the treat that was inside. Nick did this several times, after which we put the tool site inside Kanzi's enclosure. Nick knelt outside, making flakes. He handed sharp ones to me while I was inside with Kanzi, and I encouraged Kanzi to use them to cut the string. He very soon realized the utility of the sharp flake, and eagerly took it from whoever was in with him. He then quickly went to the tool site to open the box. He even knocked two rocks together on several occasions, but in a rather desultory way, and without producing flakes. Nevertheless, he was clearly emulating Nick.

During that first afternoon, and throughout the project (which is continuing), Kanzi was never required to perform a task, but merely provided with the opportunity to participate if he wanted to. We wanted to motivate him to make and use flakes, and we hoped he would learn by example. As the days and weeks passed, he became more and more determined and proficient, and he displayed a degree of persistence at the task that exceeded anything I'd seen him do. It seemed we had a potential tool-maker in our midst.

Kanzi very quickly learned to discriminate between sharp flakes and dull ones, using visual inspection and his lips. On the second afternoon, for instance, we gave him a series of ten trials, in which we provided him each time with five flakes. For the first few trials, he seemed to be working from trial and error, but in the final five he unerringly picked the sharpest flake. He developed a very keen eye for good flakes as they were being produced. On one occasion, about three weeks into the project, Rose Sevcik was striking a rock when, for the first time for her, it split and several flakes flew off in different directions. Kanzi was watching closely and seemed to know which was the best flake, even before they hit the ground. He let out a bonobo squeal of delight, rushed to pick up the sharpest flake, and was off to the tool site with it, all in one fluid motion.

Making flakes for himself, however, proved difficult. At first, he was extremely tentative in the way he hit the rocks together. Almost always he used his right hand to deliver the hammer blow. He held the core in his left hand, often cradled against his chest, or sometimes braced against the floor, with his foot adding further support. Sometimes he put the core on the ground and simply struck it with the hammerstone. No one had demonstrated this "anvil" technique to him. No matter how he held the core, however, he seemed unable or unwilling to deliver a powerful blow. Bonobos are three times stronger than a human of the same size, so there was no doubt that Kanzi had the muscle power to do the job. We wondered whether he was nervous about hitting his fingers; perhaps he lacked the correct wrist anatomy to produce a "snapping" action; or perhaps he was reluctant to deliver a hard blow, because throughout his life we had discouraged him from slamming and breaking objects.

Then, one afternoon eight weeks into the project, I was sitting in my office, which is close to the inside room where the tool site was set up. I was suddenly assailed with the sound of a BANG . . . BANG . . . BANG. It kept on and on, and I wondered what on earth had happened. I rushed to the tool-site room, and there was Kanzi, stone knapping with tremendous force. He had finally learned how to fracture rocks to make sharp flakes, albeit small ones.

During the first three months of the project Kanzi became steadily more proficient at producing flakes, in part because he seemed to have learned to aim the hammer blows at the edge of the core. But despite his willingness to deliver harder hammer blows than he had initially, he still wasn't hitting hard enough to produce flakes bigger than about an inch long. Nevertheless, he persisted with his newfound concentration, and we in turn made the string that secured the tool site thicker and thicker, so that small flakes would wear out before they cut the string.

One day during the fourth month, I was at the tool site with Kanzi, and he was having only modest success at producing flakes. He turned to me and held out the rocks, as if to say, "Here, you do it for me." He did this from time to time, and mostly I would encourage him to try some more, which is what I did that day. He just sat there looking at me, then at the rock in his hand, then at me again, apparently reflecting. I wondered what he was thinking, because he did seem to be pondering weighty matters as he gazed at the rocks. Suddenly he stood up bipedally and, with clear deliberation, threw a rock on the hard tile floor with a tremendous amount of force. The rock shattered, producing a whole shower of flakes. Kanzi vocalized ecstatically, grabbed one of the sharpest flakes, and headed for the tool site.

There was no question that Kanzi had reasoned through the problem and had found a better solution to making flakes. No one had demonstrated the efficacy of throwing. Kanzi had just worked it out for himself. I was delighted, because it demonstrated his ingenuity in the face of a difficult problem. I quickly telephoned Nick, and told him what had happened. I was so excited by the event that I didn't give a thought to the fact that Nick might not be delighted too. He wasn't. He was disappointed. "The Oldowan tool-makers used hard-hammer percussion, not throwing," he said. "If Kanzi throws the rocks, the percussion marks will be random, and we won't learn anything." Our different reactions reflected, I suppose, the different interests of the psychologist and the archeologist. Nick said I had to discourage Kanzi from throwing, and I pointed out that that would be difficult. After all, he had found an efficient method to get what he needed. "Try," said Nick. I agreed to try.

Rose Sevcik came up with the obvious suggestion, which was to cover the floor with soft carpeting. The first time Kanzi went into the carpeted room, he threw the rock a few times and looked puzzled when it didn't shatter as usual. He paused for a few seconds, looked around until he found a place where two pieces of carpet met, pulled back a piece to reveal the concrete, and hurled the rock. We have assembled a videotape of the tool-making project, which I show to scientific and more general audiences. Whenever the tape reaches this incident there is always a tremendous roar of approval as Kanzi—the hero—outwits the humans yet again.

By this time, spring was approaching, and we decided to take the tool site outdoors, where Kanzi would have to resort to hard-hammer percussion once again, as there was no hard floor to throw against outdoors. Forced to abandon his throwing technique, Kanzi steadily became more efficient at hard-hammer percussion, delivering more forceful and more precisely aimed blows. Very consistently now, Kanzi was hitting the edge of the core and was more successful at producing flakes. The resulting cores sometimes were very simple, with just a couple of flakes removed, or, if Kanzi had persistently hammered at them, they had many small flake scars and steep, battered edges, some of which resembled the "eoliths," or dawn stones, found in Europe in the decades around the turn of the century. There had been great controversy about these objects, with some arguing that they were true artifacts. They turned out to have been the product of natural forces, such as wave action or glaciation.

Just as Kanzi was becoming quite proficient at hard-hammer percussion he foiled us yet again, which again delighted the psychologist and dismayed the archeologist. Kanzi discovered that even outside on soft ground he could exploit his throwing technique. This discovery seemed to be the result of a thoughtful analysis of the problem as well: He placed a rock carefully on the ground, stepped back, and took careful aim with the second rock, poised in his right hand. His aim was true, and the rock shattered. He continued to use this technique, and there was no way of stopping him. As far as I was concerned, we had presented Kanzi with a problem and he had figured out the best way to solve it—three times.

———

Kanzi had become a tool-maker. But our question was, how good a tool-maker is he? Could he have stood shoulder to shoulder with the makers of Oldowan tools, striking flakes off cores as effectively as they did? Nick's experience as an Oldowan tool-maker offered us a way of addressing these questions.

Through teaching himself to make the apparently simple core forms and flakes of the Oldowan, Nick came to understand both the process of flaking and the product. "The mechanics of flaking stone are not intuitively obvious," he told me.[10] When he tried to teach me and some of the other Wenner-Gren conferees how to flake that afternoon on the beach in Portugal, I could understand what he meant. I was impressed by how very difficult it is to produce flakes, and the challenge is not simply to hit the core with sufficient force. You have to know where to aim the hammer blow, and how to deliver it.

The initial inclination of the naive stone knapper is to hit the core hard enough so that a flake will pop out of the core, as if it were being chiseled out. But, as Nick demonstrated, the flakes come from the bottom of the core, not the top. The best everyday example of the principle of concoidal fracture at work in stone tool-making is the effect of a tiny pebble hitting a window: A cone of glass is punched out of the pane, and the exact shape of the cone is determined by the direction at which the stone hits the glass.

For effective flaking by hard-hammer percussion, three conditions have to be met. First, the core must have an acute edge (one with an angle of less than 90°). Second, the core must be struck with a sharp, glancing blow, hitting about half an inch from the edge. And third, the blow must be directed through an area of high mass, such as a ridge or a bulge. With these conditions met, and starting with suitable raw material, one can form long, sharp flakes. Oldowan assemblages were often made from lava cobbles, which had been rounded through being carried along stream or river beds. The toolmakers sometimes struck flakes from one side of a cobble, producing the unifacial form of a chopper, or along two sides, yielding a bifacial form. Whatever forms are produced, they have the appearance of great simplicity. But as Nick correctly points out, "It is the process, not the product, that reveals the complexity of Oldowan tool-making."[11]

Nick and Kathy Schick recently drew up a list of criteria by which to assess the technological sophistication of simple cobble and flake tools. "It was necessary to get beyond relying on gut reaction for distinguishing between true artifacts and naturally fractured stone," explains Nick.[12] The criteria are as follows:

1. *Flake angle.* This is the angle formed between the striking platform of a flake and the dorsal surface of the flake (representing the edge angle of the core before the flake was detached).

2. *Degree of removal of outer unusable surface.* Many flint stones are encased in a tough concrete exterior that crumbles when it flakes and is thus unusable as a tool. In order to get to the flint, which is an inner stone that can readily be flaked to produce a sharp edge, it is first necessary to remove this rough outer core. The ratio of remaining usable to unusable material left in the stone after flaking provides an index of how efficiently the stone has been reduced by the activity of flaking. A good knapper will removed virtually all of the unusable material very easily.

3. *The size of flakes removed.* The ratio of the size of the largest flake scar to the maximum dimension of the core is a partial indication of how efficiently flakes are being detached from a block of stone.

4. *The amount of step fractures and battering seen on the stone.* Step fractures are unclean breaks in the stone. When a stone is hit at exactly the proper angle, a clean flake falls off, leaving the stone surface as smooth as if someone had run a knife through butter. When a stone is flaked at the wrong angle, it breaks off the rock, leaving many jagged edges at the point of the break. If a stone is merely slammed into another hard surface, with little regard for the angle of the blow, it may break, but it will have a battered appearance, looking precisely as though it had been battered. Well-flaked stone looks as though it had been sculpted or chiseled.[13]

Measured against these criteria, the products of the earliest tool-makers score highly. "It seems clear that early tool-making proto-humans had a good intuitive sense of the fundamentals of working stone," observes Nick.[14] They knew about angles required on the core, about sharp, glancing blows, and about seeking regions of high mass on the core. They also apparently knew when otherwise suitably shaped cobbles would not flake well.

For instance, Nick noticed that at Koobi Fora in northern Kenya, heavily weathered cobbles were common. Such cobbles flake unpredictably and inefficiently, and a knowledgeable stone knapper would avoid using them. The outward evidence of the weathering, however, is slight and appears merely as hairline fractures on the surface. When Nick examined the Oldowan assemblages at various archeological sites at Koobi Fora, he found only rare evidence of the use of weathered cobbles; their frequency at the sites was far less than their occurrence on the ancient landscape. "It seems clear that the early hominids had already learned to reject such inferior material," concludes Nick.[15]

By these measures, therefore, the Oldowan tool-makers displayed considerable technological sophistication and perceptual skills. What of Kanzi? His progress in hard-hammer percussion has been considerable, moving from the undirected, timid tapping of rocks together to the forceful hammering directed at the edge of the core. Nick describes the process of learning to make tools as being punctuational, with periods of slow change in between. "You suddenly get an insight into what is required, and then slowly improve on that," he explains.[16] Kanzi clearly had an insight into the importance of hitting the rock close to its edge; and he had important insights when he developed his throwing techniques. Despite this, however, he has not yet developed the stone-knapping skills of the Oldowan tool-makers. In a paper reporting the first eighteen months of the project, we described our assessment of Kanzi as follows:

> So far Kanzi has exhibited a relatively low degree of technological finesse in each of [the four criteria] compared to that seen in the Early Stone Age record. The amount of force he uses in hard-hammer percussion is normally less than ideal for fracturing these rocks. His flake angles when using hard-hammer percussion tend to be steep (approaching 90°), while Oldowan flakes were generally detached from more acute-edged cores (flake angles typically 75°–80°). As yet, Kanzi's cores retain a very high proportion of their original cortex and are steep-edged and rather battered. The flakes he produces tend to be relatively small (generally less than 4 cm long) and often stepped or hinged, and his cores generally exhibit marginal (non-invasive) flake scars.[17]

There is, therefore, a clear difference in the stone-knapping skills of Kanzi and the Oldowan tool-makers, which seems to imply that these early humans had indeed ceased to be apes. It isn't yet certain, however, whether Kanzi's poorer performance is the result of a cognitive or an anatomical limitation. Or simply lack of practice. Certainly most of us working with Kanzi are unable to make stone tools ourselves. Without a good teacher and constant practice it is a very difficult skill to acquire. Were an anthropologist to excavate our site a million years from now, I doubt that he or she could distinguish between the human stone artifacts and those produced by Kanzi—except, of course, those that were made by Nick. Making stone tools is not easy and does not seem to be a skill that normal human beings acquire readily with little instruction, as we are asking Kanzi to do.

Nick hopes to learn whether or not Kanzi, with minimal demonstration, can acquire a skill that took, at best, many generations for our ancestors to perfect. If Kanzi does not succeed in matching the skills of Oldowan tool-makers in the span of one research career, it would still be foolish to rule out the potential of the ape mind to do so, given a few generations of exposure to need to use such tools.

The structure of bonobos' arms, wrists, and hands is different from that of humans, and this structure constrains the ability to deliver a sharp blow by snapping the wrist, a movement that Nick considers important in effective tool-making. I suspect that if Kanzi is limited in the quality of flaking through hard-hammer percussion, it is the result of biomechanical, not cognitive, constraints. His ape hands, with long, curved fingers and short thumbs, prevent him from gripping the stones efficiently enough to allow him to deliver a powerful, glancing blow.

—

The greatest surprise of the tool-making project was Kanzi's development of throwing as a way of obtaining sharp edges. Not only did it reflect a problem-solving process in Kanzi's mind, but it also produced material that addresses an important archeological problem: What did early humans do *before* they made Oldowan tools? Application of the criteria mentioned earlier to identify genuine artifacts as compared with naturally fractured stone would reject Kanzi's flakes and cores as tools. And yet they are artifacts, and they can be used as effective cutting tools.

Some of Kanzi's cores look rather similar to Oldowan core tools, acknowledges Nick, but most do not, because of the angles and preponderance of small flake scars. "If I were surveying a Stone Age site and found some of these things, I'd definitely check them out, but I would almost certainly conclude they were

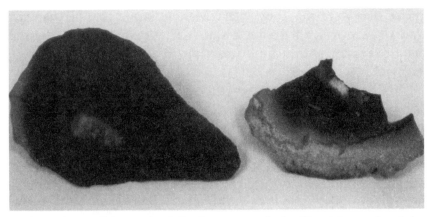

Nick Toth's tool is on the left; the one made by Kanzi is on the right. Kanzi has removed only part of the cortex, or soft material on the outside of the flint. The chopperlike shape has not been produced by design, but results from simple attempts to remove enough of the cortex to get a sharp edge. Nick's tool, by contrast, is made with a specific shape in mind and many blows have been deliberately struck to produce a tool with the shape seen here. Kanzi recognizes the value of a tool such as that made by Nick, and when given a choice between a tool he has made and a handaxe made by Nick, Kanzi chooses Nick's tools over his own, without hesitation. Occasionally, when I have left a new handaxe of Nick's in Kanzi's enclosure, I have observed him rubbing his fingers across it while looking at it very carefully. Of course I cannot know for certain, but the impression he leaves is that he is admiring the workmanship in Nick's tool.

naturally flaked," he says. But if pre-Oldowan tool-makers used Kanzi's approach, and smashed rocks by throwing them, how could an archeologist know? "Most natural processes that break rocks also tend to smooth them," explains Nick. "Rocks may fracture as they crash into each other while rolling along a stream bed, but the rolling process quickly dulls the edges."

Finding a sharp-edged cobble with angles of close to 90° might therefore be indicative of primitive tool-making. "You have to examine the context of the rock, to eliminate natural processes that might have produced sharp edges. But after seeing these incipient flaking skills with Kanzi, we certainly have to consider it as a possible model for the earliest stone tool-making. He has taught us what we should be looking for, to find tool-makers earlier than what we usually call 'the earliest tool-makers.'"[18]

When I agreed to participate in the Wenner-Gren conference I had no idea a collaboration such as the one with Nick and Kathy would arise. Both sides—in psychology and archeology—benefited tremendously. And Kanzi became the first nonhuman to learn humanlike stone tool-making in a natural setting. Nick joked that Kanzi should be awarded an honorary doctorate, pointing out that he would need a small cap and a gown with long arms. He wasn't joking, however, when in the spring of 1991, Kanzi was awarded the inaugural CRAFT Annual Award for Outstanding Research Pertaining to Human Technological Origins. Nick and Kathy are co-directors of CRAFT, or the Center for Research into the Anthropological Foundations of Technology, at Indiana University. "The award is justified, because the work with Kanzi has given us one of our most important insights into paleolithic technology," says Nick. "It has given us a view of what is possible with apes, and an insight into the cognitive background of what is necessary to go further."

— 9 —

The Origin of Language

ACCORDING to the evidence of molecular biology, the first hominid species appeared approximately five million years ago, a bipedal ape with long arms and curved fingers that presumably was well at home in the trees. The earliest known fossil evidence of such a creature dates from three to four million years ago, and was found in Ethiopia. These early hominids, with their 400-cubic-centimeter brains, and several later small-brained species, all belonged to the genus *Australopithecus*. Only when the genus *Homo* appeared did brain size begin to increase, leaping by 50 percent in *Homo habilis*, to more than 600 cubic centimeters. The next player on the stage of protoman was *Homo erectus*, who debuted almost two million years ago. His brain size varied from 850 to 1100 cubic centimeters. Modern levels of brain size—1350 cubic centimeters—came with the evolution of archaic *Homo sapiens*, probably around two hundred and fifty thousand years ago.

Much of the increase in brain size from the ape level to the modern human level can be accounted for by an enlargement of the neocortex, the thin coat of nerve cells that forms the outer layer of our brain. This outer covering did not expand with a common equipotential over the entire surface, however; the frontal lobes, associated with planning and foresight, expanded disproportionately. Another part of the brain, located in the lower rear part of the skull and termed the cerebellum, has also expanded disproportionately in man. This area is associated with the automatization of skills such as driving a car, riding a bike, buttoning a shirt, and so on. This human expansion pattern was absent in the australopithecine species, but it appeared in *Homo habilis*—or handy man—the first stone tool-maker.

It was once believed that Broca's area, long thought to be the area of the brain that made language possible, was unique to humans. Located (usually) in the left frontal lobe near the temple, Broca's area in humans is readily identified as a raised region. When evidence for Broca's area was discovered in the cranium of a 1.8 million-year-old *Homo habilis* from northern Kenya, two decades ago, it was taken as an indication of an advanced language faculty. However, we now know that Broca's area occurs in the brains of other animals, too, and is merely expanded in humans, not unique. Similarly, none of the other brain centers involved in language comprehension and production, including Wernicke's area (located in the left parietal lobe) and a scattering of some dozen or so nuclei throughout the prefrontal region, represent novel structures. The difference between ape and human brains is essentially quantitative. Those who argue for a uniquely human language-acquisition device, as do proponents of the Chomskian school, do so in the absence of any anatomical evidence for its existence.

The brain, metabolically speaking, is an extremely expensive organ. It represents 2 percent of body mass but consumes 18 percent of our energy budget. There seems little reason to have such a large brain unless it somehow greatly increases the survival prospects of its bearer. Based on the nature of size increase and reorganization seen in fossil brains, anthropologist Dean Falk, along with many others at the Wenner-Gren conference, believe that it probably was language that propelled the increase in brain size. As Dean expressed it, "If hominids weren't using and refining language, I would like to know what they *were* doing with their autocatalytically increasing brains."[1]

Terrence Deacon, a neurologist at Belmont Hospital, Massachusetts, is equally emphatic, but from the perspective of modern brains, not fossil ones: Deacon bases his conclusion on a study of the nature of differences in connectivity in ape and human brains, and on developmental studies of monkey brains. "The brain structures and circuits most altered in the course of human brain evolution reflect some unusual computational demands by natural languages," he notes.[2] These alterations center on the increasing dominance of output from the prefrontal region, which allows voluntary control over vocalizations.

The modern brain appeared with the first members of a group that is loosely called archaic *Homo sapiens*, which evolved some quarter of a million years ago. Were these people as linguistically sophisticated as we are today? It's hard to say, but if the brain-size/linguistic-capacity relationship holds, as has been argued, then the answer should be yes, for these people had brains the size of our own. They differed from us only in that they retained a physical robustness that is absent from modern skeletons.

The second line of anatomical evidence—that of the vocal apparatus—tells very much the same story as the one we see with brain-size increase. The vocal apparatus consists of the larynx (or vocal organ), the pharynx (or throat), the nasopharynx (or nasal cavity), the tongue, and the lips. In all mammals apart from humans, the larynx is positioned high in the neck, a position that has three consequences. First, the larynx can be "locked into" the nasopharynx—the air space near the "back door" of the nasal cavity. When this occurs, all breathing is done through the nose, as the back of the oral cavity is closed by the overlapping of the soft palate and the epiglottis.

Second, although the vocal tracts of chimpanzees and other mammals can produce most of the human vowel sounds, it is difficult for them to make some of the sounds readily. Edmund S. Crelin, an anatomist at the Yale University School of Medicine, has done extensive modeling of both the human and the ape vocal tracts, including the construction of manipulable rubber casts, which permit him to determine the sounds that can be produced by the physical structure of the organism. Crelin has studied this problem in detail, comparing the anatomical capacities not only of apes and humans, but also of many hominids. He has done so by reconstructing the vocal tract tissue on the basis of the available skeletal material. In order to investigate the range of sounds that a chimpanzee can make, Crelin built a rubber model of the chimpanzee vocal tract and forced pressurized air through the model as he manually manipulated its shape. These experiments led him to conclude that he could "force a rubber tract of many nonhuman mammals to produce a set of vowel-like sounds, including those of mammals with even longer snouts, such as a horse."[3] Nonetheless, he found that it required extreme constriction of the model ape vocal tract to produce the long *e* and long *u* sounds, and that it was also nearly impossible for the chimpanzee to switch rapidly between vowel sounds.

Finally, the range of noises apes can make does not include the most important element of human speech—the consonant. This is because they have difficulty accomplishing what is called velopharyngeal closure, or the brief blocking off of the nasal passages as air is forced through the mouth. This blockage is needed for the production of consonants; it enables us to generate the brief turbulence and temporary microbursts of air that are the basis of consonants. The action of the vocal cords lays noise over these temporary perturbations and the shape of the vocal tract itself is modulated to amplify or decrease certain frequencies, thereby serving to filter the action of the basic sound produced by the vibration of the vocal folds. These filtered sounds, without turbulence, become vowels; with turbulence, generated by velopharyngeal closure or a sealing off of the nasal cavity by raising the soft palate in the back of the throat very rapidly, we are able to produce interpretable speech.

Man alone has a vocal tract that permits the production of consonant sounds. These differences between our vocal tract and that of apes, while relatively minor, are significant and may be linked to the refinement of bipedal posture and the associated need to carry the head in a balanced, erect position over the center of the spine. A head with a large heavy jaw would cause its bearer to walk with a forward list and would inhibit rapid running. To achieve balanced upright posture, it was essential that the jaw structure recede and thus that the sloped vocal tract characteristic of apes become bent at a right angle. Along with the reduction of the jaw and the flattening of the face, the tongue, instead of residing entirely in the mouth, was lowered partially down into the throat to form the back of the oropharynx. The mobility of the tongue permits modulation of the oropharyngeal cavity in a manner that is not possible in the ape, whose tongue resides entirely in the mouth. Similarly, the sharp bend in the supralaryngeal airway means that the distance between the soft palate and the back of the throat is very small. By raising the soft palate, we can block off the nasal passageways, permitting us to form the turbulence necessary to create consonants.

An obvious question follows from my argument that the evolution of a bipedal mode of locomotion in our ancestors was important in the development of a vocal tract capable of producing consonants: Why did *Australopithecus* not follow the same evolutionary path as *Homo* in developing a humanlike larynx? I've referred to all species in the human family, including *Australopithecus*, as bipedal apes. This is true, in the sense of the very close genetic relationship humans have with apes. But it may be a bit misleading with respect to how efficient the different hominid species were in their bipedal locomotion.

There has been a long-running debate among anthropologists over this question, with some arguing that the australopithecine species walked just as modern humans do, that is, with a fully upright, striding gait. Others disagree, saying that the australopithecines retained many apelike adaptations, including spending a significant amount of their time in the trees and having a more shambling gait while on the ground. I support this latter argument, primarily because the australopithecines had long arms, short legs, and curved bones in their hands and feet, just as apes do. It is true that they were adapted to a degree of bipedal locomotion, but they were not fully bipedal as species of *Homo* have been, right from their first appearance two and a half million years ago. My argument over the effect of posture on the vocal tract refers to full bipedalism, not the incomplete form that prevailed in *Australopithecus*.

Of course, there also must have been changes in the neuroanatomical systems that controlled these structures. In addition to the proper anatomical design, speech production requires extremely precise and coordinated control of many muscles. Moreover, speech is so rapid that we cannot possibly be producing each sound individually. We are, instead, co-articulating, which means that our mouths have already assumed the shape for the next sound to be made before we have finished producing the first sound. Since speech is infinitely variable, the co-articulation process is never the same from one word to the next unless one repeats oneself. This means that although words sound the same to us when we hear them from one time to the next, they are not really being said in the same manner. They are altered as a function of the speech context in which they occur. It is for this reason that it is so difficult to build a device that interprets speech. Speech is infinitely varied and currently only the human ear can readily find the meaningful units in these infinitely varied patterns. The consonants permit us to accomplish this feat.

Why are consonants important? Couldn't apes and monkeys simply use the sounds that they can make to construct a language all of their own? The issue is more complicated than it seems. Studies of the vocal repertoires of chimpanzees reveals that they, like many other mammals, possess a "graded" system of vocal communication. This means that instead of producing distinct calls that can easily be distinguished from each other, they produce a set of sounds that grade into each other with no clear boundaries. In one sense, a graded system permits richer communication than a system with fixed calls, which is what characterizes many bird species, for example. The graded vocal system of chimpanzees permits them to utilize pitch, intensity, and duration to add specific affective information to their vocal signals. For example, food calls signal the degree of pleasure felt about the food, as well as the interest in food per se.

More important, these affectively loaded signals are exchanged rapidly back and forth and the parameters of pitch, intensity, and duration serve communicative functions very similar to human speech. We can say a phrase like, "Oh, I am very happy" with such feeling that the happiness almost leaps out of the speaker, or with a cynicism that lets the listener know the speaker is not really happy at all. Graded systems are well designed for transmitting emotional information that is itself graded in content. A feeling of happiness is something like a color in its endless variations. However, a word such as "fruit" or "nut" does not lend itself to a graded system. Words are units of specific information, and while they may themselves generate affect, they are not dependent on the affect for their information-bearing qualities. Consequently, unlike affective signals that constantly intergrade, words have a definite beginning and ending.

If apes indeed are intelligent enough to do so, why have they not elaborated their graded system into one with units, as we have? Unfortunately, vowels are ill-equipped to permit such "packed" communication. Even in human speech, vowels grade into one another, making it impossible to determine where one starts and ends. When tested with a computer-generated sound that slowly transitions from the vowel sound "Ah" to that of "Eh," humans exhibit a sort of "fuzzy boundary." There is a large area of the transition space that we label either as "Ah" or "Eh" without much consistency. This is true in trials of different listeners and of the same listener on different trials.

It is consonants that permit us to package vowels and therefore produce a speech stream that can be readily segmented into distinct auditory units, or word packages. Here we experience what is called a "categorical shift." If the computer presents us with consonants rather than vowels, as in a test where the sound slowly changes from "Ba" to "Pa," we continue to hear "Ba" until all of a sudden it sounds as though the computer decided to switch to "Pa." Although the computer has indeed presented us with a gray area of transition, just as it did when it played the vowels, we no longer recognize that it is happening with consonants. It is as though we have an auditory system equipped with filters designed to let us hear either "Ba" or "Pa," but nothing in between. When we hear a "Ba" it either fits the "Ba" filter parameters or it does not. If it does not fit, we cannot make a judgment about it, as we do a vowel, because we simply don't hear it as some mixture of "Ba + Pa" to judge.

For some time after this phenomenon of "categorical shift" was discovered, scientists thought that humans alone among mammals possessed the ability to process speech sounds categorically. Moreover, it was widely accepted that this capacity was a genetically predetermined aspect of our auditory system. Even though many scientists recognized that animals could learn to respond to single-word spoken commands, it was assumed that they were doing so on the basis of intonational contours, rather than the phonemic units themselves.

This view held sway until a method was devised to ask animals what they heard as they listened to consonants and vowels that graded into one another. The techniques used in these tests were modeled after those that had been applied to human infants, in which they were asked a similar question regarding their categorical skills. Human infants proved able to categorize consonants in a manner similar to that of adults, a fact that was initially viewed as strong support for the belief that these capacities were genetically programmed into our auditory systems. However, tests with mammals as different as chinchillas and rhesus monkeys revealed clearly that man was not unique in the capacity to make categorical judgments about consonants. Other animals could form acoustic boundaries that categorically differentiated consonants, even though they employed no such sounds in their own vocal systems. Thus, speech sounds are unique to humans only with regard to our ability to produce them, not with regard to our ability to hear them. On recollection, it seems odd that it should have surprised us that auditory systems are capable of far greater sound definition than the organism is able to produce with its vocal cords. After all, we live in a very noisy environment and to get along in the forest, we certainly need to be able to discriminate and make sense out of many sounds that we ourselves cannot produce.

Consonants are rather "funny" sounds. They must, for example, always be linked to a vowel if they are to be heard as a consonant. We cannot separate vowels and consonants in normal speech because it is impossible for humans to say a consonant without also saying a vowel. Thus we cannot utter "G," but rather must say "Gee" or "Ga" or "Ghuh" or some similar sound. However, it is possible, with the aid of computer, to chop apart vowels and consonants and thus have the computer say "G" in a way that we cannot. To accomplish this, you need only record some speech into a computer using a program that can transform auditory information into visual information. Once you have a picture of the sound on the screen, you can then play back the picture and watch as a time pointer moves through the wave form while listening to Ga, or any other sound you have recorded. If you watch the pointer move through the wave form while listening, you can determine the point at which you do not hear the "G" but are instead listening to the "ah." If you cut the word at this point and play the two halves, you will find something astonishing. The "ah" sounds like a normal "ah," but the "G" is not recognizable at all. It sounds like some sort of clicking, hissing noise and you will think that somehow the computer has made a mistake. But you have only to paste this hissing, clicking noise back onto the "ah" sound to hear the "G" sound again, as clear as can be.

What does this tell us? Variations in our perceptions are more a property of the auditory and neurological systems that we listen with than the sound pattern itself. Sharp, short sounds like clicks and hisses are perceived differently from tonal longer sounds. Why would this be? It may be the result of another unusual fact about clicks and hisses: We can localize them extremely well in space, a skill we probably owe to the fact that a broken branch or disturbed leaf can signal the approach of a predator. Most mammals, including ourselves, need to be able to turn in the right direction quickly and respond without hesitation when such a sound portends danger in the forest. By contrast, longer vowel-like sounds are produced by most mammalian and avian vocal tracts and are used for communicative purposes, not for hunting. We cannot localize such sounds as well. When animals hunt they are quiet, and clicks and scraping noises as they move through the forest are the only clue. Thus it seems that auditory systems have evolved different ways of listening to different sorts of sounds.

The fact that clicks and hissing sounds are so distinct and easily perceived gives them unusual properties when they are linked to vowel-like tonal sounds. The merging together of these two sound types results in what we hear as consonants. Without consonants it is doubtful that we would have spoken language. Why not? The answer is that vowel sounds are difficult to tell apart. It is hard to determine when an "ee" sound turns into an "ii" sound. At the extremes, you can determine which vowel is being produced, but this ability fails us rapidly as one vowel sound begins to grade into another.

The same is not true of short sounds like hisses and clicks. We hear them as discrete staccatolike events well localized in space. When these clicks are merged with vowels, consonants appear and act as the boundaries around vowels that permit us to determine readily where one syllable starts and another stops—so that we hear words as individual units.

It is startling to learn that these things we call words, which we hear as such distinct entities, are really not distinct at all. When we look at a visual wave form of a sentence, we find that the distinctions between words vanish completely. If we pause in our speaking for a break or for emphasis, we see a break in the wave form, but for regular speech, a sentence looks like one continuous word. Thus the units that we hear are not present in the physical energy we generate as we talk. We hear the sound spectrum of speech as segmented into words only because the consonants allow our brains to break the lump down at just the right joints—the joints we call words.

Seen from this perspective, the fascinating thing about human language becomes our ability to produce the actual units of speech. If we did not have the ability to attach clicks to vowels, we could not make consonants. Without consonants, it would be difficult to create a spoken language that could be understood, regardless of how intelligent we were.

It seems odd that the human animal is the only one that has gained the ability to produce consonants. Of course, it is also the case that we are the only animal that is a habitual biped, and the demands of bipedality have pressed some rather important constraints upon our skull. Of course, we paid some prices for these changes. Our small teeth could no longer serve as weapons, and the sharp bend in our throats left us forever prone to choking. But the ability to form consonants readily gave *Homo* a way to package vowel sounds in many different envelopes, making possible a multitude of discriminable sounds. For the first time in primate evolutionary history, it became physically possible for us to invent a language. I suspect that our intellect had the potential for language long before, but it took the serendipitous physical changes that accompanied bipedalism to permit us to package vowels and consonants together in a way that made possible the open-ended generation of discriminable sound units—the crucial step leading to speech around the world.

These unusual properties of the auditory system are paralleled by similar phenomena in the visual system. Suppose we look at a row of marquee lights flashing off and on. If the time between the flashing of light A and light B is brief enough, we will perceive the lights as a single moving piece of energy. That is, we will not see any breaks or holes in the movement; our brain will fill in the gaps. If the light is slowed down, however, we will perceive it as jumping from one marquee bulb to another, with gaps in between. Thus, at one speed we see only a moving light; at another, we see a jumping light. This visual phenomenon is, like the categorical shift phenomenon, a property of the visual system of many primates as well.

Given the perceptual constraints of the auditory system, it is evident that the appearance of language awaited the development of a vocal system capable of packaging vowel sounds with consonants. Regardless of brain size, if the vocal system of the organism could not produce consonants, it is not likely that language would emerge. The majority of land-dwelling mammals are quadrupedal and consequently have retained the sloping vocal tract designed to modify vowels, to convey affect, and to enable them to swallow easily without choking. This elongation makes rapid consonant-vowel transitions physically implausible, even if the neuro-circuitry were to permit the ape to attempt it.

What of early hominids—would their vocal tracts have permitted them to produce consonants and thus package their vowels into discriminable units of sound? Edmund Crelin has constructed model vocal tracts for *Australopithecus, Homo erectus,* Neanderthals, and other archaic *Homo sapiens.* While the reconstruction of soft tissue is always difficult, and the testing of a rubber mold is also subject to numerous subtle variations, such tests are nonetheless the current best way to approximate the speech capacities of extinct species. Crelin concluded that the ability to produce vowel-like sounds typical of modern speech would not have appeared until the advent of archaic *Homo sapiens,* around two hundred and fifty thousand years ago. These creatures had a brain capacity similar to our own.

If Crelin is correct, then language cannot have been responsible for the creation of *Homo sapiens.* Rather, it appears that gaining the vocal tract that made language possible may simply have been a free benefit as we evolved into being better bipeds. How we achieved the fine neuroanatomical control required to orchestrate the co-articulatory movements and the voluntary respiratory control to operate our vocal tract, however, remains something of a mystery.

———

At the Wenner-Gren conference several people felt uncomfortable with Kanzi's command of language comprehension. They were able to take satisfaction, however, when I acknowledged that it would probably be more difficult to teach Kanzi to tap dance than to use a keyboard-based language system. Their relief was because they felt that there might be a link between the highly developed motor skill that we use to tap dance and the similar motor-planning routines required by speech. Moreover, the idea of increased motor skill and planning as an evolutionary engine interested those, such as Patricia Greenfield, who saw evolutionary links between the development of tool-use skills and language—and suggests a common neurological substrate for both.

Evidence for the putative neurological link between tool-use skills and language includes the fact that certain brain areas, such as the inferior parietal association area, are involved in both object manipulation and language; in this case, object naming and grammar. This kind of overlap is clearly seen in certain patients who have suffered damage to their Broca's area. Depending on the nature of the damage, such people may be unable to construct grammatical sentences. In other words, they are unable to assemble words in a hierarchical manner. These same people are unable to sketch out simple hierarchical patterns, using short sticks and a model to copy. In experiments with young children, Patricia tracked the emergence of the hierarchical concept with age. She asked the children to copy a symmetrical model, again building a two-dimensional structure using short sticks. Although children aged seven and older built the structure along hierarchical lines, younger children did it piecemeal, focusing on one local area of the structure at a time.

The implication of this and other evidence is that, rather than arising in neurological isolation, as the Chomskian position argues, language abilities are intimately related early in life to abilities of object manipulation. "The ontogenesis of a tool-use program relies on Broca's area in the left hemisphere of the brain, just as early word formation does," observes Patricia. "This is the key point in relation to tools and language: they have a common neural substrate in their early ontogenetic development. . . . These programs differentiate from age two on, when Broca's area establishes differentiated circuits with the anterior prefrontal cortex."[4]

If this is indeed the case, then it is legitimate for archeologists to look for evidence of linguistic abilities encrypted in stone-tool assemblages. Changes in the cognitive sophistication of tool assemblages should—in some way—also reflect changes in linguistic capacity. The problem is, there is no objective method for assessing the degree of cognitive sophistication embodied in mute stones, as I learned from the disagreements voiced among the archeologists present at the Wenner-Gren conference in Portugal. When, almost two decades ago, Glynn Isaac tackled the challenge of looking for signs of language in tool technology, he first looked at the overall picture between

two and a half million years ago and a little more than thirty
thousand years ago. This perspective on the trajectory of language evolution through the past two and a half million years
led him to rather different conclusions from those derived from
the anatomical evidence produced by the reconstruction of
vocal tracts and observations of brain expansion and organization. Isaac concluded that the initial stage of stone-tool technology, the Oldowan, produced between two and a half and one
and a half million years ago, implied "designing and symboliz-
ing capabilities . . . not necessarily vastly beyond that of contemporary [apes]."[5]

Tom Wynn and Nick Toth have looked at the same evidence more recently, with somewhat different results. As I
indicated in Chapter 8, Tom believes the Oldowan tool technology was essentially within the cognitive reach of an ape.
"In its general features Oldowan culture was ape, not
human," he concluded. "Nowhere in this picture need we
posit elements such as language. . . ."[6] Nick, on the other
hand, states that the "Oldowan tool makers were not just
bipedal chimpanzees."[7] Nick bases his conclusions on the fact
that the earliest tool-makers apparently had mastered the principles of concoidal fracture, that is, the searching out of
appropriate angles for striking platforms and the delivering of
appropriately angled blows with a hammerstone. This mastery,
says Nick, implies a cognitive competence beyond that of apes,
and may be taken as evidence of some linguistic ability.

Nick has other evidence of the earliest tool-makers that, he
ventures, might again point to linguistic abilities. Specifically, he
discovered that in the earliest archeological sites in Kenya, the
tools were all made by right-handed individuals. This fact was
revealed by the position of pebble cortex on the flakes. Because
right-handers hold the core in their left hand and rotate it
clockwise, the cortex, when present, will mostly be on the right
edge of the flake. The opposite is true for left-handers.
Although individual apes and other primates often have a handedness preference for manipulative tasks, as a species they are
equally divided between right-handers and left-handers.
Humans are unusual in having a population preference for one

hand—90 percent of humans are right-handed. Handedness is associated with localization of function to the opposite brain hemisphere. The location of manipulative skills in the left hemispheres of (most) right-handers is accompanied by the location there of language skills, too. The right hemisphere has become specialized for spatial skills.

If it is true that language and manipulative skills evolved together, it seems likely that brain lateralization was part of that process. In that case, perhaps the Oldowan tool-makers, being preferentially right-handed, could have increased linguistic skills after all? The asymmetry of brain function is accompanied in modern humans by an asymmetry of shape: The left hemisphere is emphasized. This same asymmetry is seen in the earliest known crania of *Homo habilis*, the putative first tool-maker, but not in any australopithecine species. Nick sees this as further evidence of linguistic abilities in the Oldowan tool-makers.

Despite their differences of interpretation over the Oldowan evidence, Nick and Tom agree over the general significance of the next stage, the Acheulian. Unlike the simple core tools of the Oldowan, which may be formed by the removal of a very few flakes, the Acheulian handaxe involves extensive flaking and about fifteen minutes of effort. "Besides manifesting a clearer sense of spatial geometry [than in Oldowan tools], the technical sophistication of [handaxes] is such that even modern humans learning to make stone tools often require many months of apprenticeship to reach the requisite level of finesse," observes Nick.[8] This is evidence for cultural norms, he suggests, and a greater language ability than in the Oldowan tool-makers.

Tom sees "something humanlike in this product of *Homo erectus* minds,"[9] referring to the handaxe. "Artifacts such as these indicate that the shape of the final product *was* a concern of the knapper and that we can use this intention as a tiny window into the mind of *Homo erectus*."[10]

These three lines of anatomical evidence—of the brain, the vocal apparatus, and the capacity for tool-use—provide the principal support for the notion of long, gradual changes on the road to language. Along with these changes in the brain and the vocal apparatus, there occurred concomitant gradual changes in the hand, changes that made it an increasingly suitable instrument for tool construction and use. As we saw in Chapter 8, Kanzi has difficulty flaking and using stone tools because of his elongated fingers, which cannot be fully straightened, his inflexible wrist, and his short and highly positioned thumb. While he can see how we hold the cobble and the hammerstone, and he can attempt to imitate our grip, he can never position the stones as effectively as we can for flaking.

Nonetheless, the coincident changes in the anatomy of the hand and wrist, and the increasing signs of tool construction and use in the fossil record have led many scholars to conclude that a common neurological structure undergirds both skills and that this structure is to be found only in *Homo*. To say that a common neurological substrate may exist still does not explain the nature of the driving force that was pressing brain expansion at such an unprecedented rate in all of mammalian development. Yes, *Homo habilis* was making simple tools and *Australopithecus* probably was not, but did *Homo habilis* require a brain that was 50 percent again larger than that of *Australopithecus*? Given Kanzi's ability to flake stone, the answer obviously seems to be no. What then was pushing the brain to become larger?

It may be that actually making a tool is far simpler than having the foresight to know that a tool would be necessary in the future. Making a tool entails a great deal more than simply bashing stone together. In order to make a tool, one must have the right sort of stone at hand and one must not be engaged in another activity. In Kanzi's case, these conditions are provided; we bring the rock and we focus his attention on the need for a tool by baiting the tool site. However, we were not around to help out *Homo habilis*. He had to plan ahead. He had to know that at some future time he would need a tool, even though he did not need it for a specific task at the time he found the proper stone or at the time he actually constructed the tool. Thus, the real cognitive demands placed on any tool-user entail finding the proper material, making the tool, and keeping the tool with him or her until it is needed. And all of this must happen without a stimulus, that is, without some immediately present set of circumstances that make the need for a tool manifest. If a tool-maker recognized that he needed a tool only when the occasion actually arose, it is likely that (1) there would be no appropriate stone around, and (2) by the time he made the tool, the situation leading to its need would have changed.

Consequently, it is not simply an understanding of geometry that a tool-maker must have; first and foremost the tool-maker must have the ability and the intelligence to plan ahead, to disjoin present behavior from future need. For this, the tool-maker must construct an elaborate mental model of the anticipated future. This model must include the tool and its purpose as well as the tool's shape and the actions required to produce that shape. Once the tool is made, the maker must keep the tool in his or her possession until it is needed. This may seem simple, but when there are streams to be crossed, berries to be picked, predators to avoid, children to watch, and when the tool must be on hand at night as well, keeping this tool at ready access is not a trivial task. Only to the extent of anticipating future need, in the absence of any current reminder, will a hominid go to the trouble of carrying and keeping a single tool. Keeping a kit of such stone implements becomes even more difficult. Indeed, once the stone tool kit becomes larger than two or three implements, it is easy to see why temporary home bases, scattered throughout a range, would become necessary and convenient for storage.

Many animals, including apes, sea otters, and birds, use tools—but only apes and man make tools, and only man makes tools that he is not going to use in the immediate future. In addition, man carries tools long distances for days or weeks anticipating their future use. Man keeps tools with him, through daylight and darkness, through bad weather and good. Unlike other animals, man searches far and wide for good raw material from which to make his tools and then transports that material to other locations. All of these skills require considerable foresight and planning, skills attributed to the frontal lobes. They require constant orientation to the future and a clear understanding that present and future needs may not be the same. The greater the ability to plan ahead, anticipate future needs, and prepare for them, the greater the chance of survival for a creature who has small teeth and minimal means of natural defense.

———

One day Nick showed me a tape of some of the last surviving stone tool manufacturers, who live in a remote section of the New Guinea highlands. These men learned their skills as youths, before their people had experienced any contact with the outside world. It was clearly an activity they enjoyed and were quite proficient at. They talked and sang as they worked, sometimes hardly even watching their own handiwork, it was so second nature to them. The knowledge of how to construct the tools seemed to reside in their hands rather than in their head. Their heads were busy spinning tales of past and future occasions where tools had been and would be needed.

The 50 percent jump in brain size between *Australopithecus* and *Homo habilis* seems to reflect the increased need to plan ahead that makes stone tool production possible. As man began to plan further and further ahead, he surely constructed increasingly elaborate mental models of what might happen in the future. At some point, he inevitably began to construct multiple models, each with different plausible scenarios and each requiring a different sort of preparation. At that point, man could no longer coordinate his behavior effectively with glances, gestures, emotions, and common knowledge of the situation. To coordinate and select between different mental models of the future, man needed language. With language he could describe his model of the anticipated future to others. This holding in mind, and contrasting of multiple models, must have required an increase in cortical capacity. However, the actual making of the tools and even the syntactical and phonemic skills used to describe the various mental models are, and probably were, rather automatized and once acquired, needed little additional cranial space. Potential mental models of the future are not the kinds of ideas that readily become automatized. They tend to remain in conscious awareness and to be constantly updated by the events at hand.

Somewhere between five hundred thousand and fifty thousand years ago, a hominid species became human, as *Homo erectus* gave way to *Homo sapiens*. At first these *Homo sapiens* were archaic in form, but by one hundred thousand years ago, they began to merge into the form we know today. During the same time, technology progressed from the simplest manufacture and use of stone flakes and cores to the production of elaborate implements that required a clear mental concept, or template, and considerable manual dexterity. Ultimately, some tools took on an aesthetic appearance, too, in the form of beautifully fashioned and decorated implements made of stone, bone, and antler. About thirty thousand years ago, during the period archeologists call the Upper Paleolithic, artistic expression became manifested, in the form of body ornamentation, images painted and engraved on walls, and the carving of bone and ivory figures. Compared with archeological evidence of human behavior earlier than thirty thousand years ago, the advent of the Upper Paleolithic was a dramatic and explosive event, and it included the practice of elaborate rituals, the establishment of larger communities, and the development of long-range contacts, possibly involving trade. Methods of hunting became much more efficient, and people began to settle in the far reaches of the world, including Australia and the Americas. Revolution is not too strong a word to describe the magnitude and rate of change that took place.

There is no doubt that the nature of human behavior we see in the Upper Paleolithic was an expression of the modern human mind. However, there is currently little agreement as to degree to whether these behavioral changes were biologically or culturally driven. The puzzle archeologists face is this: If anatomically modern humans evolved a hundred thousand years ago, why did modern human behavior not appear until some seventy thousand years later? The fact that no obvious anatomical change corresponded to what can only be called the dawn of cultural life has led many to ponder the place of language in this set of events.

Was it language that made all of this possible? Iain Davidson certainly thought so, and he was the first to have had the patience to explain the cultural explosion in such detail to me that I began to grasp the full weight of its importance.

To Iain, *Homo erectus* had barely even taken the first step on the journey toward modern human behavior, let alone progressed some distance on it. "We propose that all human ancestors without language should be considered as closer to chimpanzees than to modern humans in their behavior," Iain wrote in a paper he co-authored with William Noble for the Cascais conference volume. "Two events in the record of prehistoric evolution of human behavior can be said to be the first that unambiguously entail the existence of language: the colonization of Australia, before 40,000 years ago, by people crossing the sea to an unknown shore; and the appearance of sculptures and bas reliefs with coded symbols in different parts of Europe before 32,000 years ago."[11]

Iain and his colleague have developed a hypothesis that argues that before our ancestors created images, not only *did they not* have language, but they *could not* have had language. They acknowledge that the art of Upper Paleolithic Europe betrays the existence of full-blown symbolic language, and suggest that earlier, more primitive image-making activities must have preceded it. These earlier stages would have marked the emergence of the referential abilities that characterize image making and symbolic language. True language evolved sixty thousand years ago, or perhaps a little earlier, suggests Iain. Prior to that, our ancestors' vocal skills were mere "context specific communications."[12]

Iain and Bill's argument is twofold. First, it dismisses the suggestion that the Acheulian handaxe implies intentionality on the part of the tool-maker and therefore the inferences of cognitive and linguistic skills that go with it. They claim, instead, that the characteristic shape of the handaxe is the incidental outcome of removing many flakes from a particularly large core, not the deliberate fashioning of a shape according to a mental template. And the consistency of shape through space and time is the result of "a small number of learned motor actions,"[13] not the systematic imposition of arbitrary form. They point out that the sticks made by chimpanzees to catch termites conform to local regularity, and that this conformity should be interpreted as a consequence of the chimps having a small repertoire of actions for preparing the tool.

"If handaxes were produced according to a mental template, that would indeed imply considerable intentionality," Iain observes. "The hominid would be sitting there for fully ten or fifteen minutes, working steadily toward a planned goal, and that is not a trivial intellectual exercise. But the paradox of the archeological record is that there is absolutely no other evidence of this putative level of intentionality. If that competence were present, you would see evidence of it," concludes Iain.[14]

Although Nick Toth concedes that Iain has a good point, he rejects the notion that Acheulian handaxes are the byproducts of repetitive flaking. "It makes no sense to me," he says. "If you do what he says, you finish up with a discoidal core, not a tear-drop shaped handaxe."[15]

When Iain was visiting the Language Research Center early in 1993, I thought I would perform a small test of his hypothesis, and so challenged him to produce a handaxe by knocking flakes off a large core. He had to be guided only by the need to remove flakes efficiently, I charged, and must not intentionally produce the characteristic shape of the Acheulian axe. He failed. Perhaps the test was unfair, because flaking is a considerable skill in itself, and Iain was not fully adept at it. My view is that Iain is probably wrong on this point, and that *Homo erectus* stone knappers knew what they were aiming for when they made handaxes.

The second line of argument in Iain and Bill's hypothesis relates to the nature of language itself. The essence of symbolic language is its referential nature, the act of invoking through the use of symbols an event or object in its absence. Words come to replace, or refer to, those actions or objects in the minds of speaker and listener. The ability to displace reference to an object from the presence of the object itself was engendered by the act of using images, Iain and Bill argue. The scenario is something like the following.

An individual points to a bison and perhaps vocalizes but, critically, makes a simple iconic image in the soil or on a wall, perhaps reflecting the curve of the head and back. Later, that same image invokes the notion of bison, but in the absence of the animal; and so might the associated vocalization. The crucial mediator in the process, insist Iain and Bill, is the production of the visual image that separates the object from a reference to it. "Depiction . . . provokes the reflectivity that in turn permits referential utterance," they wrote in a scientific paper.[16] "Communication is common to many creatures, but only humans have the capacity to communicate their meanings independent of context."[17]

If Iain is correct and language as we know it appeared on the scene only forty thousand years ago, this means that large-brained *Homo sapiens* existed on this planet for nearly sixty thousand years without language. It means that in the forty thousand years since we invented language we have come to depend on it so completely that we now believe it to be innate. It also means that *Homo erectus*, *Homo habilis*, and *Australopithecus* could not have had anything like language. I finally began to see why it was that Iain was wary of the data I had assembled on Kanzi. I also began to wonder what it could have been like to be a member of a human group in which people communicated by nonverbal expressions and gestures, but did not use language. If Iain is correct, what were we like before we invented language? I thought of those vague references to "dreamtime" people in aboriginal cultures, and the references in our own culture to the absence of "knowledge of good and evil" before Eve consumed the proverbial apple. I also recalled those references to some African and Indian cultures in which it is said that older brother and younger brother decided upon different paths long ago when they first became aware that it was possible to control fire. It is said that the older brother elected to remain in the forest, following the old ways and eschewing fire and language. The apes of today are descended from older brother. Younger brother went out from the forest and kept fire with him, becoming the progenitor of all humans today. Could cultural myths such as these hark back to a murky time in our distant past when we possessed human minds but no language?

Though I greatly admired Iain's work, it was difficult to accept his view that language was a recent visitor on the evolutionary playing field. If this were true, how could Kanzi understand complex, novel sentences like, "Can you go scare Matata with the snake?" I knew my data were sound. Iain had a similar confidence in his data. Moreover, we each respected the care and critical eye that the other brought to the process of reconstructing the roots that undergird human existence. Could such diverse perspectives meet?

Language arose in a particular manner in human prehistory, and this process must have impressed itself in a consistent way in the evidence available today. Read correctly, these lines of evidence, together with those from modern neurology and psychology, should therefore tell the same evolutionary story. Yet two extreme views of language origins have emerged from this data. The differences that Iain and I shared were symptomatic of the disparate opinions regarding the role of language in human evolution.

The existence of such divergent views on the origin of language suggests that, in some way, the archeological and fossil evidence is being misread. Although my initial motivation for working with apes was to help in the search for a greater understanding of human behavior, I have been unquestionably affected by the data unfolding over the past decades. These recent studies have become increasingly relevant to the unresolved question of the evolution of human language. The data provided by the abilities of Sherman and Austin, coupled with the data of Kanzi and Panbanisha, provide an independent means to evaluate the two models of language origins.

While what we have learned about the linguistic and tool-constructing capacities of apes does not tell us *how* hominids became human, it does tell us a great deal about the common substrate of mind shared by apes and our hominid ancestors. Our modern minds evolved from these ancestral minds, which must have shared many characteristics with apes not too different from those alive today. Thus Kanzi and other apes offer a glimpse of the starting point in man's evolutionary transformation from a state of nature to the modern human condition.

———

In his second most famous book, *The Descent of Man*, Charles

Darwin noted the great paradox of the gap between human achievement and that of the rest of the animal world. "[T]here is no fundamental difference between man and the higher mammals in their mental faculties," he observed. And yet, he acknowledged, "Of the high importance of the intellectual faculties there can be no doubt, for man mainly owes to them his predominant position in the world."[18]

What precisely are these higher "mental faculties"? Every standard text on human evolution cites culture and language. Culture is defined as the process of passing knowledge from one generation to the next, and language is defined as the vehicle by which that knowledge is passed. But what of that period between one hundred thousand years ago, when *Homo sapiens* appeared with his large brain and his fully upright posture, and forty thousand years ago, when the "cultural explosion" took place?

Were we passing knowledge down from generation to generation during this period? If so, regardless of whether or not we had invented language, the knowledge we were transmitting must have been very different from the kind of political, ritual, and technical knowledge that we pass along to new generations today. Certainly the tool manufacturing techniques were simple. Language was neither necessary nor probably helpful in passing along the methods of tool construction. Modern tool-makers tend to pass their knowledge along by example, a point made convincingly by both Patricia Greenfield and Tom Wynn at the Wenner-Gren conference. Patricia studied the cultural transmission of weaving in a South American culture and found very little use of language in this activity. Tom, reviewing the literature on the transmission of tool construction and use, concluded that "tool behavior is learned largely through apprenticeship—each actor constructs his own constellations. They are not shared."

Indeed, the same can be said of Kanzi. Nick did not utilize language in showing Kanzi how to flake stone, but rather treated him as an apprentice. Kanzi observed, but constructed his own constellation of techniques as well. Thus it would not seem to be the case that we need language to pass along major dimensions of culture. In fact, a moment's reflection regarding the kinds of behavior that travel easily from culture to culture suggests that a great deal of information passes from culture to culture without language. In Japan, baseball is extremely popular, jeans sell for outrageous prices, and music has taken on a distinctly Western flavor. All of the changes permitted by modern technology travel rapidly around the world, yet without a common language. Try to watch a Japanese baseball game and figure out what the letters on the screen mean or what the announcer is saying. Languages are frozen bits of culture compared to most other sorts of behavior. Languages are clung to, much like other behaviors that fall under the label of "ritual." When people move from one culture to another they bring with them their language, their religion, their marriage ceremonies, their rites of passage, their child-rearing patterns and their kinship patterns. They acquire and readily adapt to the new forms of shelter, the new foods, the new sports, and the new tools. Thus we see that one of the main uses of language is to transmit the practice of language itself.

Surely archaic *Homo sapiens* of two hundred and fifty thousand years ago were capable of passing along techniques of tool use and construction, artistic rendering, shelter construction, and so on, if they knew how to do such things. The fact that they left no mark of such knowledge in the archeological record suggests not they were incapable of passing along these capacities to the next generation, with or without language, but that they had not yet developed these skills. Clearly, early *Homo sapiens* led an existence so different from that of even the most primitive hunting and gathering societies today, that it is not easy to contemplate what that existence was like.

I therefore agree with Iain, and others, who suggest that language might have been a cultural invention, as agriculture was. The organizational and technological activities associated with food production and, later, the building of city states, far outdistance those required for a hunting and gathering society. And yet it was the brain of a hunting and gathering species—*Homo sapiens*—that mastered these new skills. No genetic mutation occurred that allowed people, ten thousand years ago, to do what their forebears did not. Similarly with language, it is quite conceivable that the cognitive apparatus that underlies language might have evolved for other, related purposes, such as planning future actions, tool construction, or social negotiation. In any case, the fact that language-mediating centers are scattered throughout much of the prefrontal region of the brain, and that their location may differ from individual to individual, indicates that a large learning element is present in language acquisition, perhaps exclusively so.

I do not agree with Iain in his contention that a capacity for symbolic communication must necessarily postdate image making. The central message of the ape-language work is clear. Not only can apes acquire a capacity for symbolic communication through the structured experience we developed with Sherman and Austin, but also they can acquire it spontaneously through casual exposure to language, just as human infants do. It has been easy for Kanzi to comprehend language; the key was in his early exposure to speech and the pairing of the speech with symbols. His comprehension skills are so extensive as to make it implausible to offer a conditioning account. His younger sisters, Mulika and Panbanisha, have followed in his footsteps, revealing that it is the rearing experiences Kanzi received, not Kanzi himself, that permitted him to understand simple language. If Kanzi can acquire language so readily, we must conclude that the ape brain is capable of a primitive language. Either bonobos are utilizing this capability in the wild in ways we have not yet grasped, or exposure to the invention of language is all that Kanzi and his family needed to bootstrap their way into language.

As we saw earlier, spoken language owes its unprecedented information-carrying capacity not just to the large range of sounds produced by the unique human vocal apparatus, but also to the nature of some of those sounds. The ability to produce consonants in association with vowels, therefore, led the way to the production of an extremely large number of discernible sounds. Kanzi's comprehension of human speech suggests that he is able to decode consonants. If he could produce consonants as well as vowels, and if he had the requisite degree of neurological control, I have little doubt that Kanzi would be able to speak.

The ability to produce spoken, symbolic language depended, therefore, on the appropriate development of the vocal tract in early human ancestors, not on the evolution of the requisite cognitive capacity. Even in primitive form, such a system of communication would have had considerable survival advantages; and the sophistication of the system would have increased through time as natural selection honed those advantages. Why this occurred in the *Homo* lineage and not in the australopithecines—as indicated by brain size and organization, and vocal tract anatomy—is an important question. Almost certainly this capacity for spoken language was associated with a more intense social nexus, connected with a more complex subsistence strategy—the beginnings of what eventually blossomed into a hunting and gathering way of life.

Our work with Sherman and Austin, and particularly with Kanzi, has offered an independent way of judging the interpretation of language origins based on more traditional lines of evidence. The unexpected cognitive underpinnings of symbolic language possessed by modern apes suggests that the early members of the human family were equipped with a greater intellectual capacity for language than is usually assumed.

———

If early *Homo sapiens* were indeed without language, each of us must feel compelled to ask: What were we like, we large-brained, sensitive, intelligent creatures producing limited technology and no art? We couldn't have been even remotely similar to any living people extant today. For example, even a cursory look at the Sambia stone-age people of New Guinea reveals that language is critical to the entire structure of their culture, their initiation rites, their sexual behavior, their diet, everything they do. Not that they use language to *teach* these activities; rather, they employ language to proscribe them, to set the rules for when they occur, how they occur, why they occur, and what will happen if they do not occur in the proper way at the proper time.

Young males, for example, enter into exclusive homosexuality with older males prior to the onset of adolescence. They maintain these relationships, without the knowledge of the female members of the group, until well into their twenties, when they take a wife and become exclusively heterosexual. It is language that imposes, guides, and directs this channeling of sexual orientation. It is also language that defines, guides, and directs the development of fighting skills. That is, language is not used to teach a boy how to fight, but it does define with whom he should fight and why, as well as the level of prowess to be obtained in interaction with enemies. Indeed, it is language that defines the identity of the enemy and that carries grudges against the enemy across one generation to the next, even though younger members of the group may never have experienced in person a reason to define a particular group as friend or foe.

Without language would we carry such vendettas across generations? Without language would we feel a need to define possessions as belonging to one person or another? Without language, how far ahead would we understand cause-effect relationships? Would we understand the relationship between sexuality and childbirth? Without language would we construct elaborate kinship structures and the obligations that accompany them? Without language would we have recognized ourselves as independent beings, responsible for our own actions? Would we have had any need of moral codes of right and wrong? Would we have had need of clothing for the sake of privacy? Would we have had any understanding of death?

— 10 —

At the Brink of the Human Mind

IN an essay he wrote almost two decades ago, the Rockefeller University philosopher Thomas Nagel posed the question, "What is it like to be a bat?" The point of the essay was to explore the problem of gaining insight into another individual's mind, and in particular the mind of another species. Nagel was prompted into the exercise by the work of fellow Rockefeller University scholar Donald Griffin, who, against the accepted scientific wisdom of our time, argued that animals other than humans engaged in reasoning about their world; in other words, said Griffin, humans are not the only species to experience the phenomenon of mind.

Nagel's answer to this question was, "We can never know." He maintained that the bat's perceptual world is so different from ours, based as it is on echolocation of high-frequency sounds beyond our experience, that we can never comprehend the bat's mental world. The Austrian philosopher Ludwig Wittgenstein framed the conundrum in a different way, but with the same conclusion. "If a lion could talk," he said, "we would not understand him." Customs and cultural conventions create barriers of understanding between different human societies, observed Wittgenstein, so imagine how much more difficult it must be to penetrate the mental life of another species.

I suspect that the problem has been exaggerated, for several reasons, not the least of which is the uncertainty over the phenomenon of mind itself. We saw in the first chapter of this book that Western philosophy and science have grappled with the issue, from Aristotle, who viewed humans as uniquely endowed with a rational soul, through Descartes, to whom only humans possessed a rational mind, to the modern behaviorists, who, like Descartes, restrict mind solely to *Homo sapiens*. Despite Griffin's persistence in encouraging biologists and psychologists to buck the behaviorists' hegemony and consider the existence of cognitive processes in the nonhuman domain, there remains a hesitancy to accept the idea of minds in species other than our own.

As noted in Chapter 1, the concept of mind as we human beings experience it—that is, mediated by symbolic language and carrying a cogent essence of self-awareness—has for many people come to represent an unbreachable boundary between humans and nonhumans. Man stands safely on his side of the wall, declaring his separateness from the rest of nature. In part he has done so because the societies he has constructed have isolated animals from his daily existence. Animals no longer share the forest or the plains with us and only through the medium of television can we be impressed by the ways in which they cope with life. In most zoos and laboratories, it is difficult for animals to engage in behavior that displays intelligence, as their food is provided for them and their social groups are temporary, making it difficult, if not impossible, for them to establish traditions.

When we take animals such as dogs into our homes, they often seem to do things that appear intelligent, but scientists are taught, early in their career, to beware of any interpretations that smack of anthropomorphism—the fear is that the intelligence may be in the eyes of the observer rather than in the mind of the animal. By labeling most interesting descriptions of animal behavior as anthropomorphic anecdotes, we unwittingly eliminate the need for serious scientific attempts to understand such behavior. That is, by assuming that the complex and interesting behavior is in the mind of the observer rather than the animal, it becomes the human observer rather than the animal that we seek to understand.

But perhaps the deeper reason we so readily declare our uniqueness from animals is to assure ourselves that we are indeed reasoning creatures with a culture created by our own hand and mind. By setting ourselves apart from animals, we experience some small measure of safety. If we *Homo sapiens* are truly different by virtue of reason, we can look to reason to protect us from falling into the trap of reacting instinctively and losing the evolutionary game. When we look at animals we do not see cities or villages or agriculture or possessions—and their way of life does not look like that to which we would aspire, no matter how much we enjoy watching movies of wildlife. Therefore, it is comforting to assume that we have reason and culture by nature and that these abilities will always keep us from returning to any sort of animal state.

But at the expense of gaining some comfort in our ability to plan our future, we risk alienating ourselves psychologically from all of the other creatures on this planet. After all, we have all evolved together. As a species, we are just beginning to understand that it is our view of separateness that has led us blindly to exploit the world of nature, be it by destroying natural habitats or performing experiments on animals with insufficient concern for the effect of these actions either upon the animals or upon ourselves as moral beings.

It is becoming increasingly evident, not only from work with apes, but from studies of species as varied as dolphins, parrots, sea lions, elephants, and wolves, that man has deluded himself by focusing on this separateness. As we come to understand other animals better, our current notion of human uniqueness will likely change and we will realize that future generations may view us as having looked at animals through a distorted lens, much as we now look back at the early explorers who thought that different races of man reflected different levels of evolution.

To recognize the connection between our intellect and mind and those of other creatures on our common planet, we must permit ourselves to ask questions in a careful and serious

way about behaviors that reflect animal intelligence in complex situations. As we begin to do so, we will have to alter our
focus from the test paradigms that closet our thoughts. These
closet paradigms are experimentally correct, but their very
structure limits the behavioral options animals may exhibit to
an array of simple responses. Such limitations, carried out for
the purpose of upholding the scientific method and prevent-
ing self-deception, have unwittingly forced the egregiously
erroneous conclusion that animals show little capacity for
what we call inference, insight, reasoning, or thought.

These closet paradigms came about as post-Darwinian
inquiries into behavior sought to follow the models set by
physics and chemistry. But behavior can never be successfully
understood using these models. Living organisms, as subjects of
treatments, are not the same from one test to another. Each
action upon a living organism alters it in some manner, making
it impossible to recreate the former conditions. Researchers
have tried to get around this fact by repeating the stimulus presentation until the behavior is stable. Unfortunately, this technique serves only to create the illusion of control over behavior.
In fact, it manufactures an artificial situation that limits the
options available to the animal and thereby causes it to appear
to be under the control of the stimulus. A more fundamental
problem arises when researchers attempt to follow the models
of physics and chemistry by concentrating on the antecedents of
one or two behaviors or even a set of replicable behaviors. Such
attempts assume that behaviors can be mixed and matched
something like a set of chemical elements, and that an orderly
and predictable reaction will follow. Unlike chemicals, behaviors
cannot be reasonably separated from the entire context in which
they occur. That context encompasses both the actions of the
animal across time and the events within the environment across
time. Indeed, behavior is fundamentally a time-based phenomenon.

Behavior is nothing more than, and nothing less than,
changes in patterns of action that take place only in the domain
of time. Chemical reactions by contrast, though they may
require time, are transmutations of inanimate compounds in the

spatial domain. That is, they are structural alterations that manifest themselves on a physical plane. Behaviors manifest themselves on the temporal plane. While they may require chemical alterations within the nervous system to drive them, the patterns of behavior are neither explainable nor fully predictable by those chemical reactions—for they are always in a constant state of flux as responses to the changing environment that surrounds the organism.

For most organisms with a complex nervous system, the world is experienced as a constantly altering series of events. Though some events may seem similar to previous experiences, in the real world, unlike the laboratory, they are never really the same. The organism must constantly adjust to a new state of affairs and select the behavioral options that are most appropriate. Decisions that put certain options into play will themselves alter the range of options available in the immediate and distant future. All organisms with complex nervous systems are faced with the moment-by-moment question that is posed by life: What shall I do next?

The world out there is never precisely the same from one moment to the next; consequently, the behavior that occurs in response to perceived events can never be the same, nor should it be, if the organism is functioning in a normal manner. Only when behavior becomes abnormal do organisms engage in repetitive patterns of behavior that do not take into account the constant environmental changes around them. When we see this happen, we call such patterns stereotypies and recognize that they are abnormal. Yet, in our attempts to follow the models of physics and chemistry, many laboratory studies of behavior unwittingly create behaviors that are analogous to stereotypies. In our attempts to present the same stimulus conditions across repeated trials, we destroy the very phenomenon we set out to study.

As long as behavioral scientists follow in the footsteps of Descartes, assuming that nonhuman animals are merely robots made of meat and bone, they will refuse to give up their paradigms built upon the methods of physics and chemistry.

Using these models, they will continue to come up with

experimentally solid and verifiable scientific data to support their initial hypothesis. Indeed, their very goal will be to design experiments to support these hypotheses. In so doing, behavioral scientists do not, as did Einstein, look and wonder at physical phenomena out there in the world. By the design of their laboratory and their apparatus, they inadvertently create the animals' physical world, thereby limiting the potential responses even before they frame their hypothesis. It would be as if Einstein had designed space and time and, having done so, arrived at the theory of relativity and then decided to test it.

————

The physical world that surrounds us is of a different order from the animal world. The greater the degree of development of the nervous system, the more these two worlds differ in kind. The purpose of complex nervous systems is to permit flexible actions, unique to each situation. To the extent that organisms take in environmental information and make decisions about future actions, they become increasingly different from inanimate matter and require different paradigms for their study.

Once organisms have developed nervous systems sufficiently complex to postulate presumed goals and/or intentions for other living creatures, a world based on the moment-by-moment interpretations of the intent of others will arise. That is, it will be the presumed intent behind the behavior, rather than the specific actions themselves, that will mold the response of the observer. Certainly, humans and apes have entered the world of interpreted intent, and I suspect that a number of other animals have done so as well.

Some students of animal behavior have sought to escape the limitations posed by the laboratory by engaging in fieldwork. They rightly argue that by observing an animal in its natural habitat, one does not arbitrarily constrain the range of options available to it. The view that the behavior of animals is fundamentally different from that of man nonetheless manages to hold sway, even among those who do field observations. This is, in part, because a special language has been devised to describe and label behaviors of animals in nature. This devised language carefully avoids using any terms that we would apply to similar human behaviors. The taboo against using terms reserved for humans to describe the behavior of animals becomes most apparent when we observe apes. If a chimpanzee frowns and presses his lips together in a display of anger, either feigned or real, fieldworkers do not say he was mad; they say he displayed a bulged lips face. When he smiles and hugs another animal after a fisticuffs, they say he displayed an open-mouth bared-teeth grin and an arm around. Field researchers are admonished to speak in this manner in order to avoid the bugbear of anthropomorphism—the act of attributing human emotions to animals.

Consequently, out of fear that we might see a humanlike emotion where it does not exist, we design ways of speaking that permit us to talk about animal behavior without attributing any humanlike emotion to the animals whatsoever. We therefore approach the study of the animal mind with the unwritten assumption that it would be an error of the greatest magnitude if we were to conclude wrongly, in any circumstance, that an animal (even one that shares 99 percent of our DNA) felt as we do when angry or happy. Thus, even when we observe the animal in nature, the way we are taught as scientists to ask our questions, to structure our data, and to discuss what we see, constrains the conclusions we permit ourselves to find.

What if we were to assume at least partial continuity of emotional expression and intelligence between animals and man and thus permit ourselves to talk about animal behavior in a new light? We might risk the error of sometimes attributing capacities that did not exist, but we would surely find humanlike capacities that do exist but are currently hidden from us by the blinders we press over our eyes. Would the error of sometimes erroneously attributing capacities that did not exist be greater than that of never discovering any emotional or intellectual capacities that were continuous with our own? I think not. At the very least, if one scientist made a mistake and attributed some capacity to an animal that was far beyond its true ability, another scientist would come along and correct this mistake. As the situation currently stands, we don't even have the right to make the mistake. We should be able to ask questions about how animals perceive their worlds, their roles in those worlds, and what kinds of events or relationships alter these perceptions. It seems to me that this is what a science of animal behavior should be about.

I am not the first to suggest that we need to look at animal behavior through a different lens. Donald Griffin has been urging us to do so since the 1970s. Why are so many behavioral scientists still unable to break out of their constraints? The problem lies in part in the omniscience we attribute to the human mind, an omniscience that we believe is made possible by the gift of language. Even students of behavior who would not deny minds to animals nonetheless maintain there is no way of gaining access to the animal mind; consequently, they believe the issue is, like that of religious experience, beyond science. I suggest that we expand our ways of doing science to encompass such questions.

———

The essence of the difference between the human and the animal mind is often claimed to be that man can reflect upon his actions while animals, lacking words, cannot. Crucial to this view is the underlying and unspoken premise that language is the only possible means of reflection. Without language how can we ask an animal what it is thinking? And without language how can it tell us? And if it cannot tell us, how can we legitimately assume that it is thinking?

Once, while riding in the car through the woods with Kanzi's sister Panbanisha, I noted she appeared to be very quiet and pensive. I was moved to ask her what she was thinking—a question I generally avoid since I have no means of validating the answer, nor even of determining if an ape understands the question. Occasionally, when I have posed this question in the past, I have generally been ignored. However, at this moment Panbanisha looked literally lost in thought, and so I dared. She seemed to reflect upon the question a few seconds and then answered "Kanzi." I was very surprised, as she almost never uses Kanzi's name. I replied, "Oh, you are thinking about Kanzi, are you?" and she vocalized in agreement, "Whuh, whuh, whuh."

Similarly, one time I was riding in a car with Heather, one of the normal children in our project, through the very same area of woods. Heather was two years of age and just beginning to form sentences. She, like Panbanisha, typically ignored questions like "What are you thinking about?" But, like Panbanisha, at this moment she appeared lost in thought and so I dared to inquire. She replied "Mommy." I asked, "Do you wish your Mommy was here?" and she nodded her head.

I cannot be certain that either Panbanisha or Heather was really thinking. Currently there is no way to establish scientific consensus regarding the inner thoughts of another person. Yet it seems that credence should be given to the fact that both Heather and Panbanisha, on occasions when they appeared pensive, elected to answer the question. On other occasions, when they were engaged in other activities, the question was ignored as though it were nonsensical. These observations suggest that it is possible that children and apes think in a reflexive sense, even before they are competent language users. Could it be that they think in some way other than with words?

Thought, or the manipulation of one's mental model of the world, surely must take place in the absence of language, utilizing neurological machinery that services the channels of perception through which the world is viewed. It requires but a moment's reflection to recognize that humans engaged in complex nonverbal activities—such as in dance, music, sculpting, and athletic skills—depend on wordless thought. To suggest otherwise "is a notion that only a college professor or other professional wordsmith could have ever taken seriously."[1]

Mary Midgely, the British philosopher of science, puts the issue more generally: "If language were really the only source of conceptual order, all animals except man would live in a totally disordered world. They could not be said to vary in intelligence, since they could not have the use of anything that could reasonably be called intelligence at all. . . . The truth seems to be that—even for humans—a great deal of the order in the world is pre-verbally determined, being the gift of faculties we share with other animals."[2] Nonhuman animals quite evidently live in ordered worlds, an outcome of their own cognitive processes. Without such mental ordering, the management of the myriad interactions among other members of a community and the efficient exploitation of a diverse resource base would be nearly impossible. There is no question that language enhances thought processes, permitting a more intricate and powerful manipulation of mental worlds. But this is surely an extension of faculties already in place, not the establishment of something novel. Spoken language, and the thoughts it mediates, is built on the same neurological foundation that underlies thinking in nonhuman animals.

The apes I know behave every living, breathing moment as though they have minds that are much like my own. They may not think about as many things, or in the depth that I do, and they may not plan as far ahead as I do. Apes make tools and coordinate their actions during the hunting of prey, such as monkeys. But no ape has been observed to plan far enough ahead to combine the skills of tool construction and hunting for a common purpose. Such activities were a prime factor in the lives of early hominids. These greater skills that I have as a human being are the reason that I am able to construct my own shelter, earn my own salary, and follow written laws. They allow me to behave as a civilized person but they do not mean that I *think* while apes merely *react*.

Although I gain a fuller understanding of the thoughts of apes when they elect to use the keyboard, it is possible, even without words, to perceive much of what they are thinking. More important, one does not need to be especially intuitive or insightful to do so. One simply needs to be observant of their behaviors and receptive of their communications, while recognizing that these communications take into account our common knowledge of the surrounding events—a sort of joint awareness that leads to joint perception and joint knowing.

This sort of joint understanding of the world around us would be as difficult with a species whose perceptual capacities are unlike our own as Thomas Nagel argues. Dolphins can inspect space by sound, seeing things underwater that are not apparent to us. Similarly, dogs can hear and smell things that are invisible to us. Therefore it is difficult for us to make sense of many of their actions, as we do not experience the influx of information from the world as they do. However, if we did experience the world in the same manner, it seems to me likely that the processes we call "making sense of that information" would not be too dissimilar.

Fortunately for those of us working with apes, they sense the world much as we do. Their vision, hearing, sense of smell, and so on are all very much like our own. Consequently, what gains my attention is often the same as what gains theirs. For example, when walking in the woods with both chimps and dogs, I once spied a deer from 100 meters when we were in an area where the vegetation had sporadic clearings. When Panbanisha saw me look intently in the direction of the deer, she followed my glance and immediately saw the deer. The dogs, however, did not follow my glance and thus did not see the deer. Because Panbanisha and I readily and quickly follow each other's glances and because our visual systems are similarly constructed, she and I developed a joint knowledge of the deer's presence that was shared at once by us, but not by the dogs. Once she had seen the deer, she met my glance to determine my reaction.

There are also times when the dogs perceive things that elude both myself and Panbanisha. One evening we were walking in the woods at dusk when both dogs suddenly growled fiercely and turned toward something just off the trail. The dogs did not glance at either of us, though they did glance at each other. Neither Panbanisha nor I had heard or seen anything that was alarming. Straining to look in the direction of the dog's orientation, we both dimly made out the shadowy outline of a large feline perched on the branch of a tree. I do not know how the dogs discerned the presence of the big cat, but I do know that Panbanisha saw it just as I did. We immediately looked at each other and her hair stood out three inches around her body; mine did also. No words were needed for us to understand what the other had seen, nor to share the other's sense of apprehension. It also seemed obvious to both of us that the direction to head was back to the lab. We needed no words.

Once we returned to the laboratory, Kanzi, Matata, and Panzee seemed aware that something had frightened us. They also inferred that it had happened to us outside in the woods from whence we had just come. This was apparent because after taking one look at us they strained to look out into the darkness and made the soft "whu-uh" sounds that signal something unusual. Panbanisha vocalized toward them, as if to tell them about the big cat we saw in the woods. Matata, Kanzi, Tamuli, Panzee, and Neema all listened and responded in kind with very loud vocalizations of their own. Did she tell them something in sounds I could not decode? I don't know. I then relayed the story in my own way, with spoken language, to Kanzi and Panzee, as I knew they could understand something of what I said. Both of them listened with rapt attention and huge round eyes. At appropriate points during my recounting Panbanisha embellished my tale with bonobo "Waa" vocalizations, as though to add her own emphasis.

Did they understand what was said or where this had happened to us? I cannot be certain, of course, but both Kanzi and Panzee displayed hesitation and fear in that precise area of the woods the next time they were permitted to go out. Since they had never been frightened there before, it seemed that they must have understood something of what had happened.

Of course, one can conclude little on the basis of this situation alone; however, there are many others, each different. Only by looking at a large number of such situations can one begin to understand whether apes are capable of communicating such complex information. For it is not just a single event such as this that suggests communications of complex information are being achieved, but many other events, each unique and impossible to replicate without sacrificing the novelty of the setting and hence the impetus for communication.

A very different sort of example occurred one afternoon as I was playing with Matata and her daughter Tamuli, who knows no language. Tamuli asked for my keys by pointing to them and looking at my face with a questioning expression. She then played with them for perhaps thirty minutes before dropping them and becoming interested in some toys I had brought for her. When I was ready to leave I forgot to retrieve my keys. As soon as I walked out of Tamuli's room, Kanzi, who was in an adjacent area, asked to visit Matata and Tamuli himself.

I was about to unlock the door between their rooms when I realized I could not do so because I did not have my keys. I turned to Tamuli and asked her to look for the keys, not really thinking that she would understand or cooperate. To my surprise she set about at once looking under all the toys and blankets to try to find my keys. When she discovered them, she rushed over and showed them to me, but refused to let me have them back. I coaxed and cajoled for fifteen minutes, offering to trade many prized items for them, but to no avail. I could see that she was not going to give back the keys.

Finally I turned to Kanzi and explained that I was unable to open his door because Tamuli would not return my keys. It then occurred to me that perhaps I should solicit his help as he often seemed to be able to communicate with Tamuli far better than I could. "Please tell Tamuli to give me my keys," I implored. Kanzi climbed to the top of his room where wire mesh separated his area from Tamuli and looked at her while making a small noise. Tamuli approached Kanzi, looking directly at him. Kanzi made several multisyllabic sounds to Tamuli. Tamuli listened, then to my amazement quietly walked over and handed my keys back.

Did Kanzi tell her to give me my keys? Did she understand him and comply? It certainly seemed so. If such an event were to occur between human siblings, we would call it language, even if it were in a tongue we could not yet recognize or catalogue.

To further our understanding of animal intelligence we must learn to ask better questions—questions that focus on unusual events, rather than mundane and readily controllable ones. If we were to start with the assumption that animals are conscious and capable of thought, reason, and complex communication, we would find it difficult to come up with evidence that would completely disprove this view. Instead, we start with the premise that they are incapable of such accomplishments and find it difficult to disprove this view.

We do not realize how deeply our starting assumptions affect the way we go about looking for and interpreting the data we collect. We should recognize that nonhuman organisms need not meet every new definition of human language, tool use, mind, or consciousness in order to have versions of their own that are worthy of serious study. We have set ourselves too much apart, grasping for definitions that will distinguish man from all other life on the planet. We must rejoin the great stream of life from whence we arose and strive to see within it the seeds of all we are and all we may become.

Our definitions of man, readied anew for each additional discovery of capacities in animals, continue to impede our sense of belonging to the greater whole. In demeaning the capacities of animals, we found it easy to glorify our own. Having invented language, we turned and looked down upon the well-spring of life from which we arose with something akin to disdain. We catalogued our achievements, chronicling in detail how distinct they were from all other creatures, hesitant even to say that a continuum existed between ourselves and them. Not only did we deny animals the potential for thought, we assumed they had no awareness of their own existences.

Gordon Gallup, a psychologist at the State University of New York, Albany, two decades ago performed what became a classic experiment, as it was the first attempt at a behavioral definition of self-awareness. The experiment was conceptually simple, but not as easy to perform as it might seem. Gallup determined whether chimpanzees were able to recognize their own images in a mirror by placing a red, odorless/tasteless dot on their foreheads while they were anesthetized. Once the chimpanzees awoke they paid no attention to the dot until they happened to glance in the mirror. Then they noticed the dot and immediately set about removing it.

Chimpanzees, like ourselves, are quite concerned about appearances and show mirror recognition early, sometimes even before human children. Panzee has been the most precocious of our apes in this regard, passing her version of the dot test accidentally at six months of age when she first became interested in her reflection. I noticed that when a mirror was present in the room, every so often she would walk by it and check out her reflection. But I was uncertain as to whether she recognized herself or not. One day, while playing, she nicked her browridge ever so slightly, but enough that a tiny red scrape appeared. At the time, she paid this mild bump no heed whatsoever, continuing her rough and rowdy play as though nothing had happened. About forty-five minutes after this injury, having not once paid attention to the minor scrape in the meantime, Panzee happened to be walking by the mirror and, as usual, stopped to glance at her reflection. This time, however, she paused, sat down, and gazed intently for about forty-five seconds. Then she slowly reached up and touched the red spot on her forehead with her index finger while watching herself in the mirror. After touching the spot, she then looked at her finger, but there was no blood, as the scrape was so slight. She then leaned forward and looked at the spot more closely in the mirror for about fifteen seconds. Then, seemingly having satisfied her curiosity, she moved on to other activities and paid the scrape no more attention at all.

None of our other apes have displayed mirror recognition at such an early age; however, they have all demonstrated, in a variety of ways, that they not only understand their own image, but they enjoy playing with it and even altering it. Moreover, they recognize themselves not only in mirrors, but on television as well. We first observed this one day when we were testing Austin's sorting skill while taping the session. Austin was sorting colored blocks and sipping an orange drink at the time he began to notice himself in the live monitor I had set up. We could tell that he was playing with his television image because he kept bobbing up and down and making funny faces while staring at the monitor from only a few inches away. He then positioned his face adjacent to the screen and began to sip his orange drink, carefully observing himself roll the liquid from one side of his mouth to the other.

Food manipulation with the lips is something that apes are very good at and spend a great deal of time practicing. Since, in the wild, their hands are often occupied holding onto tree limbs, they have to be able to pick and peel fruit with their lips and teeth. They become very adroit with their lips and seem to take great pride in this skill. I once observed Sherman demonstrate for me his ability to roll a dime, standing on edge, from one side of his mouth to the other and back.

Austin's initial fascination with his video image engaged his undivided attention for twenty minutes. Later on, he hit upon the idea of using the television to look down his throat. He had found this difficult to do with a mirror, since when he tilted his head back he could not look very far down his throat. He soon realized, however, that the television, unlike the mirror, could show him his throat from a variety of different angles depending on the position of the camera. Consequently, he began to open his mouth extremely wide while looking at the television screen so that he could see what was at the bottom of his throat. We would then point the camera directly down into his mouth while he watched on television. This was Austin's idea, not mine. Earlier, when he first got a good look at his throat and found it too dark to see, he left to look for a flashlight and then returned to the same position to shine the light down his throat. Such an activity was never suggested to him, nor even demonstrated for him. It emerged from his own desire to see what the bottom of his throat looked like.

Sherman had no interest in this at all, and though he watched Austin look at his throat on television many times, he never did so himself. Sherman had other uses for the television image. He preferred to use it for brushing his teeth, applying lipstick, and putting large fur shawls around his shoulders while bobbing up and down—apparently to practice the fearsomeness of his display. Even though Austin watched Sherman do all of these things, he elected to do none of them. He looked at his throat and either watched himself eat or, if he had no food, pretended that he was eating out of an empty bowl with an empty spoon, studiously observing his image as he did so.

These were the things that engaged Austin and Sherman. They were individually specific and their separate interests emerged from within each of them. If we attempted to elicit these behaviors on purpose we were almost never successful. They appeared only when the apes themselves were interested and engaged with the television.

Kanzi was different yet again. He neither brushed his teeth nor looked down his throat. Instead, he used his image to practice blowing up balloons and blowing bubbles with bubble gum. He wanted to be good at both things and both were difficult for him. None of the other apes used a mirror for these purposes. Panbanisha did not really use the mirror in the sense that the others did. She was content to admire herself in the mirror, checking her appearance from every angle. When her canines began to come in, she spent long periods of time simply looking at these new large and potent teeth in the mirror. She also, at this time, began to use them.

Sherman and Austin not only recognized themselves in the mirror and on television, they also differentiated between live and taped images of themselves. When they saw a self-image, they would stick out their tongues, bob their heads, and so on to see if it were a live image. If it was live, the behaviors I described above would begin to appear. If the image did not react, they typically went on about their business, ignoring the television unless something very interesting was happening on the screen.

Working with Emil Menzel, we were even able to test this differentiation in a controlled way by using a port through which the chimps extended their hand. A television camera was focused on their arm and presented them a picture of what their arm was doing on the other side of the port. Once they put their hand through the port, the only way they could see what it was doing was by watching the television. They were able to perform all sorts of manipulative feats that required visual monitoring of the hand. For example, if presented with two closed clear containers, one with food and one without, they reached for the one with food. Even when the television image was reversed, they still utilized it, albeit a bit awkwardly, to guide their hand movements. When the live image of their hand was replaced, without warning, with a taped image of their hand, they quickly recognized that the movements they were making did not correspond with what they saw on the television (even though it was their own hand) and withdrew their hand from the port, since they preferred not to have it extended if they could not visually monitor what it was doing.

Most chimpanzees pass Gallup's dot test easily once they have had sufficient exposure to mirrors, though early isolation rearing can cause them to fail, suggesting some disruption of self-concept as a result of lack of socialization. Monkeys, however, are notorious for failing this test in spite of many attempts, regardless of rearing conditions. The curious thing is that monkeys do learn how mirrors work; when they see someone behind them in the mirror, they turn around to look, rather than attempting to touch the object in the mirror. It has even been found that when lightweight plastic daisy flowers are attached to collars on their necks, monkeys will notice the flowers in the mirror and reach behind their heads to remove them.

Thus the puzzle remains as to why they do not pass the dot test. Are they unaware that the image in the mirror is really them? Do they a lack a self-concept? Can it be that they understand everything they see in the mirror, except who they are? These results are often used to support such a view by philosophers and scientists who view self-concept as something that either must be present in the form that most human adults experience it, or entirely absent.

It could be that monkeys simply do not care that they have a dot on their fur. In the wild, chimpanzees and baboons have been observed feeding together in the same tree on sticky-gooey fruit. When wads of the fruit get stuck in the chimpanzees' hair, they stop eating to remove them when they are clearly visible on thighs or stomachs. Baboons, however, seem oblivious to these sticky wads even though they can clearly see them. Perhaps monkeys are more dependent on other monkeys for self-tidiness than apes, or perhaps their self-concept does not deal as much with appearances as does the self-concept of apes.

If we are to explore the psychological worlds of other animals we would benefit from accepting the fact that they may have self-concepts that are not equivalent to our own in form or content. In fact, it is far more rewarding to press to understand minds and self-concepts that are not equivalent in all manners to our own.

To fail to try to understand the world from the point of view of the lion or the bat is to admit that the human existence is so limited that it cannot project itself satisfactorily into the minds of different creatures. Do we really want to accept this limitation when we quite satisfactorily project ourselves into all sorts of invented imaginary creatures, even those with very different sensory systems and value systems than our own? All one need do is to read a few comic books to conclude that the projection abilities of our species are great indeed and that our children, at least, have little difficulty in going beyond their ordinary frameworks of reality.

Even within our own culture, the way individuals react to their own image depends, to a large extent, on how they feel about themselves. On one occasion I was videotaping some of the children with impairments in Mary Ann's project, to document their keyboard communication skills. One child noted himself on the live monitor I was using and grew interested in his image. Consequently, I turned the monitor around so that all the children could see themselves and continued to tape. One by one, I focused on the different children. Many of them seemed puzzled, never having seen themselves on live television before. Most of them stared at the television for some time. After about ten minutes several of them winked, stuck out their tongues, or waved their hands to see if that was really them on the television. Others grew very shy and turned away from the image as though bothered every time the camera zoomed in on them. Still others glanced at themselves furtively every so often. The ones that showed clear self-recognition by sticking out their tongues, for example, were also the ones that were the most outgoing—the showoffs of the group. The ones that turned away from their own image were the shyest of the students. Shall we conclude that they did not recognize themselves because they ignored their image?

Dogs fail Gallup's dot test also. Must we infer that they also lack a self-concept? If they don't have a concept of self, what do they have when they look and listen and carry out our commands? What kind of a concept do they have when one dog tells another that he is not to touch his bone, even when he is out of the room? The dot test is an important milestone, but unfortunately it has prevented us from asking other questions about what it is that is happening when an animal looks into a mirror.

Dogs sometimes stare into mirrors, as though attempting to make sense of the scene there. On one occasion I approached a dog in the process of studiously gazing at its reflection in the mirror. I stood quietly behind the dog, next to an open door that was also reflected in the mirror. The dog continued to gaze at itself with studied intensity. After a few moments I began to move my hands, and immediately, the dog's gaze shifted from looking at its own mirror image to mine. As it watched me, its ears became erect as if listening for the slightest sound. I moved my hands in larger motions and suddenly the dog began to bark in alarm at my mirror image. When it was clear that the dog was quite upset, I spoke. As soon as I spoke the dog wheeled around and ran straight to me, wagging its whole body in a desire to be patted. I patted the dog until it calmed down.

As soon as it was calm, it went directly back to the mirror of its own accord and assumed the identical location it had been in before and began to study the images in the mirror once again. I stood in the same place, approximately six feet behind and a foot to the left of the dog. The dog's eyes looked first at its own image and then looked again for mine, and once it found it, began again to stare intently. As I again stood very still, the dog made no sound. After approximately thirty seconds, I once more moved my hands and immediately the dog began to bark at my image with great distress. A little later I spoke and again the dog turned quickly and rushed to me to be patted.

The dog returned to the mirror and once again took the same position and this time immediately began looking for my image. I moved and was no longer reflected in the mirror. After gazing in the mirror as though searching for me but not finding me, the dog turned and walked away.

It was hard to avoid the impression that the dog was attempting to understand how there could be two of me, or how I could appear to be in two places at the same time. At the very least the dog was attempting to puzzle its way through something as it returned twice to search for what it had previously seen in the mirror.

Much of what we wish to learn about animals and their minds can be learned only when they are already engaged with the question we wish to pursue at the time. We cannot place a mirror in front of a dog, put a dot on its head, and then wait for self-recognition. The dog must be engaged with the mirror and we currently do not understand dogs well enough to bring this engagement to the forefront at will.

Engagement with a topic of interest is a widely accepted phenomenon in our own species—though one not well understood. There are children who become engaged with chess, for example, and learn all the rules easily, though they may do poorly in school in many subjects. There are children who become engaged with dinosaurs and can tell you the names of a hundred different species though they cannot recall a single date in their history class. There are three-year-olds who become engaged with backhoes and tractors and can tell a John Deere from a Ford tractor, though they cannot remember a nursery rhyme they have heard forty times.

Engagement, and its relationship to the accumulation and processing of information, is a little-studied phenomenon, representing as it does, individual skills rather than those that can be measured in a group of people.

Currently, our understanding and measurement of human intellectual capacity is oriented toward group skills and toward activities that can be elicited on command, regardless of the state of engagement. Indeed, being able to engage one's focus on the questions of the examiner, rather than on one's own interest, is the primary measure of test-taking ability, and test-taking ability is the primary measure of intelligence. When we find that animals do not do well when compared to people in this way, we must not assume that we have really measured their intellect. Perhaps we have measured only our own limited ability to engage them.

———

As researchers have become more interested in self-awareness they have begun to recognize that it is closely linked to other-awareness and that language skills are predicated on awareness of both self and others. Why would anyone bother to tell anyone else something unless they assumed that the other person did not already know what it was they were about to say?

This idea—that the knowledge states of the speaker and the listener can in fact be different—has been given the term "theory of mind." How can we know if animals believe that other animals have minds? Richard Byrne and Andrew Whiten of the University of Saint Andrews have suggested that the key lies in the ability to deceive. If one animal sets about to deceive another with intent, it must recognize that the other has a mind that is composed of knowledge states different from its own.

Lies are notorious in the animal kingdom. Nonpoisonous butterflies imitate the wing patterns of the monarch butterfly, so as to avoid being eaten by birds who have learned to leave the poisonous monarch alone. Plovers pretend to have a broken wing in order to lead potential predators away from their young. While it is clear that the butterfly that imitates the monarch is not doing so intentionally, it is less clear with the plover.

Scientists have assumed that the broken wing display is an innate pattern, carried out with little understanding in time of danger. However, this view has recently been challenged by Carolyn Ristau of Rockefeller University, who points out that like all supposedly innate patterns, the bird appears to be making decisions that vary from instance to instance and that do take into account characteristics of the predator. There are also great differences among plovers in how, where, and when they elect to lead predators, and some are far better deceivers than others. Yet none of these facts tells us whether the plover knows that it is deceiving the predator. The only time that plovers appear to have broken wings, however, is when there is a predator present.

Other animals engage in deceptive acts that are no less situation-specific. Byrne and Whiten have coined the term "tactical deception" to separate acts that are deceptive but apparently without reasoned intent, from those where the intent of the perpetrator is clearly to mislead. They define tactical deception as "an individual's capacity to use an 'honest act' from his normal repertoire in a different context, such that even familiar individuals are misled."[3]

Byrne and Whiten became interested in deception after seeing several instances of it among a troop of baboons they were observing in the Drakensberg Mountains of southern Africa. For instance, one day they saw Paul, a juvenile male, approach Mel, a mature female, who was engaged in unearthing a succulent tuber. Paul looked around: No other baboons were in sight, but they were not far away. Suddenly, Paul let out a piercing scream, as if he were in danger. Predictably, Paul's mother, who was dominant to Mel, rushed to the scene and drove Mel away. Paul then calmly ate the abandoned tuber. "Watching the incident, it was difficult to suppress an intentional interpretation," Byrne and Whiten commented later.[4] Had Paul concluded, "If I scream, my mother will assume Mel is attacking me; she'll run to defend me; and I will be left with the juicy tuber to eat"? or had he merely been upset that he was not getting any of the tuber for himself and screamed out of frustration?

The work of Dorothy Cheney and Robert Seyfarth with vervet monkeys in the Amboseli National Park in Kenya represents a long-term attempt to understand the minds of monkeys in their natural habitat. They too, have seen behaviors that appear to be intentional deception, though they have concluded that the monkeys only partially understand what they are doing. As an example, they cite the behavior of Kitui, a low-ranking male at Amboseli. One day a new male seemed poised to join the group from elsewhere, an event that would surely have been a threat to Kitui's already low social status. When he saw the male, Kitui gave a leopard-alarm call even though there was no leopard in sight. The call had the effect of keeping the new male in the trees and delaying his entry into the group. "So far, so good," observed Cheney and Seyfarth. "The alarm calls appeared deceitful because they signal danger that Kitui, but not the interloper, knew to be false, and they kept the interloper temporarily at bay."[5] But Kitui's behavior was not consistent. Instead of remaining in his own tree throughout, which is what vervets do when a genuine leopard alarm is made, Kitui descended, crossed open ground, and climbed a tree near to the new male, alarm-calling all the time. He had got only part of the story right, not unlike the child who vigorously denies raiding the cookie jar, with crumbs evident all over her face.

Overall, Cheney and Seyfarth concluded that "vervets' cognitive abilities are limited compared with our own and that there is no evidence that the monkeys have a 'theory of mind' that allows them to recognize their own knowledge and attribute mental states to others."[6] Another illustration of this conclusion is the vervet's failure to take account of the listeners' state of knowledge or ignorance when making alarm calls: The caller is uninfluenced by how much others know about what danger is threatening.

I have observed similar incidents involving Kanzi's mother, Matata. Once, when I was introducing a new person to Matata, she became jealous of the new person and refused to let her touch any item that she was fond of, including her blankets, her bowls, her food, and her mirror. One day we were sitting together on the floor when Matata decided to ask me to go get some food by holding her empty bowl out to me and making a food sound. I told Matata I would get her some food and left the room, leaving the bowl with her. I had been gone less than five minutes when suddenly Matata began screaming loudly. When I rushed back into the room the new person was holding Matata's bowl and Matata was screaming at her and threatening to bite. Matata looked back and forth from me to this person and then to the bowl, screaming—intimating that the bowl had been grabbed from her in my absence and that I should support her in attacking this mean individual who had taken her bowl. Of course, given the gift of language, this person was able to explain that she had indeed done nothing. Matata had placed the bowl in her hands and then starting screaming for me as if she had been wronged.

When Matata saw us talking about what had happened, she began to look very crestfallen, concluding her ruse had not worked. She stopped screaming and moved to the corner where she suddenly became very preoccupied with grooming herself.

When Byrne and Whiten talked informally with fellow psychologists and primatologists, they heard similar reports. Few such reports make it into the pages of the scientific literature, however, because they are single observations that the researchers have no means of replicating. In behavioral science, one is supposed to be able to predict that a certain behavior is going to occur. Clearly, when one is dealing with deception, this is virtually impossible, since one almost never knows when deception is going to occur. In fact, if the deception is truly successful, the researcher as well as the other animals are so satisfactorily deceived that they do not know that it even occurred.

Byrne and Whiten conducted a survey of more than a hundred of their colleagues in 1985 and again in 1989, asking for anecdotal accounts of incidents that might be judged as tactical deception. The response was enthusiastic, and several hundred putative cases of deception were assembled. The question was: Did these cases reflect instances of deception with reasoned intent, or could there be other explanations? When Byrne and Whiten applied strict criteria to the supposed examples of deception, ruling out as carefully as they could possibilities of learning, they concluded that of the 253 cases assembled in the 1989 survey, only 16 could be said to reflect mind reading in the sense of tactical deception. All of these were in apes, and most with chimpanzees. Two are noted here.

During her lifelong monumental studies at Gombe National Park, in Tanzania, Jane Goodall observed many examples of putative deception. One of the most intriguing involved the chimp named Figan, who noted that a banana had been left in the crook of a tree. Unfortunately for Figan, Goliath, the Alpha male, was resting under that very tree. After glancing briefly at the fruit and then at Goliath, Figan moved some distance away, perhaps fearing that if the fruit were still in his line of sight, he wouldn't be able to resist looking at it and would thus alert Goliath to its presence. After fifteen minutes Goliath left, and without hesitating Figan retrieved the banana and ate it quickly.

The Dutch primatologist Frans Plooij observed a similar incident, also at Gombe. An adult male was alone in a feeding area when a box was opened electronically, revealing the presence of bananas. Just then a second chimp arrived, and the first one quickly closed the box and ambled off nonchalantly, looking as if nothing unusual were afoot. Like Figan before him, he waited until the intruder departed and then quickly opened the box to retrieve the bananas. Unlike Figan, however, he had been tricked. The other chimp had not left but had hidden, waiting to see what was going on. The deceiver had been deceived.

———

Evidence of self-awareness and of deception therefore suggests that apes think of themselves and others as having knowledge states that differ. Is there another window into the animal consciousness, one that perhaps requires less inference regarding intent on the part of the observer? One area that has begun to be studied is that of animal pretense or imagination.

All animals, like all children, love to play, and in play all of the elements of true aggression are acted out in pretense. It is as though all of what might happen, or could happen, is experienced many times on many planes prior to the actual occurrence, so that the animal or person is ready for the event when it actually does occur. Animal behaviorists have typically assumed that animals do not know they are pretending at aggression, as children do when they play at being superheroes. This is because children can say they are pretending while animals cannot. Many researchers even claim that it is not until four years of age that children can discuss the difference between pretense and reality. Prior to this time, what appears to be pretense to us is said to be reality to the child.

Games of pretend among apes are not as elaborate as those seen in children, but they are engaged in with enthusiasm nonetheless. Vickie, the first chimpanzee who participated in a language project, was reported to have a great time pulling an imaginary toy with an imaginary string. On occasion, it even appeared that the imaginary toy became stuck, such as between the toilet and the wall, as Vickie ran around the room with it.

When he was young, Austin often pretended to eat imaginary food, occasionally even using an imaginary dish and an imaginary spoon. He would carefully place the nonexistent food in his mouth and then roll it around on his lips, watching it just as though it were real food. Sherman was not interested in imaginary food—he always wanted the real thing—but he loved to pretend that dolls, particularly King Kong dolls, were biting his fingers and toys as well as having fights with each other.

Most frequently, these pretend games were played alone, but not always. On one occasion, as Sherman and Austin were watching a King Kong movie on television, they began to pretend that King Kong was actually in a cage located in their room. This cage looked just like the one that Kong was in in the movie. At this point, Sherman and Austin stopped watching the television and began to make threat barks at the empty cage and to throw things at it, as though it housed Kong himself. Sherman even got out the hose and began to spray the cage.

As a youngster, Kanzi liked to pretend that he was hiding imaginary food in his blankets or under some toys. Occasionally he would pretend to give Panbanisha or others bites of the imaginary food. If he elected to eat it, he did not do so the way Austin did, who consumed it very slowly while watching himself in the mirror. Instead, Kanzi pretended to gulp it down hurriedly, as if he had stolen it.

Panbanisha's favorite pretend game was to act as though she had heard a monster in the next room. Going toward the door with her hair out, she would comment "monster" and invite others to search with her. Sometimes she would then put on a monster mask and pretend to chase her sister Tamuli. She also liked to pretend that she was taking bites out of pictures of food she saw in magazines, or even out of the peaches that are depicted on all the Georgia license plates at the laboratory.

But whether it is self-awareness, awareness of the minds of others, pretense, or deception—all of these cognitive activities are manifest in language, for it is with language that Kanzi and Panbanisha and Sherman and Austin can tell us things that we would otherwise not know. Kanzi has told us where he left a ball the day before and reminded us of yesterday's promise that we forgot. Panbanisha has told us that she wanted to watch an "ice TV" when it started snowing outside. Kanzi told us that he was looking for his mother, Matata, when asked why he was trying to crawl under the railroad ties. Panbanisha tore the "good" lexigram off the keyboard and gave it to us as a way of sealing her promise to be good. Sherman told me there was a "scare" outdoors when he saw a chimp being carried away in a transport cage, and Austin always tells me he wants Coke instead of the juice I am offering. Panbanisha tells me when it is raining outdoors. She also told me that the lady who visited has hair that looks like a mushroom—and she was really right. These are small things perhaps, but they offer a constant glimpse of other minds that I would not have without language.

Kanzi and Panbanisha tell me and others these things because they assume that I do not know them. Things we both know they never bother to state. Not only do they use language to present their thoughts, sometimes they use language to pretend and sometimes to lie. Kanzi, for example, knowing that he cannot have any more M&M's, will ask to go play in the T-room and get some toys when he (but not I) knows that M&M's have been placed in a T-room cabinet. Once in the T-room, he will quickly steal the M&M's and run out. He does this so deftly and quickly, it is clear that it was his intent all along.

———

Kanzi's naturalistic acquisition of words and his emergent comprehension of complex spoken sentences indicate that the chimpanzee has all the basic neurological machinery for a primitive language. Kanzi has not learned to speak, but this limitation appears to be one of the physical structure of vocal/respiratory circuitry and anatomy. His expression of a language facility through being steeped in language, as human children are, illuminates the nature of the chimpanzee mind—and it puts the mind of *Homo sapiens* in proper biological perspective.

Kanzi's linguistic capacities give us a clearer view of human language. First, they demonstrate the narrowness of the Chomskian assertion that spoken language is an evolutionary novelty that arose uniquely in humans. Second, they emphasize the interactional aspects of language acquisition, for both apes and humans. Steeped in a language-rich environment, human children first come to comprehend language, and then to produce it, a process requiring elaborate control over the vocal/respiratory tract and complex planning of muscle movement.

As language is learned, many general cognitive processes become shaped as a consequence. Language in humans, rather than being innate, is seen as the product of a plastic cognitive substrate interacting with environmental exposure to speech. In their natural state, chimpanzees do not develop the kind of language we see in captivity, but given the appropriate environmental exposure, an ability for symbolic language use becomes evident. At the same time other cognitive skills become honed. For instance, language-competent chimpanzees develop an ability to learn to use a joystick (connected to a computer monitor) through simple observation rather than through active teaching. Language-naive apes must be trained to do this bit by bit.

Clearly, exposure to speech in infancy shapes the developmental processes in the ape brain just as it does in the human brain. And it is more than likely that the neural networks influenced by a language-steeped environment in apes are evolutionary parallels to those involved in language acquisition in human children. Because chimpanzees are so closely related to humans in an evolutionary sense, it should not be surprising—except to a Chomskian—that they are sensitive to the same environments that foster cognitive competence in human children.

The ease with which Kanzi acquired a facility for symbolic communication not only tells us something about humans, and the assumed uniqueness of the human mind, but also something about apes and their cognitive competence in their natural state. We have yet to appreciate the intellectual challenge of the natural lives of intensely social primate communities. One aspect of this is that the natural communicative skills of chimpanzees in the wild are almost certainly greatly underestimated.

———

The boundary wall between humans and apes has finally been breached. As a result, more and more biologists are bowing to the logic of the evidence of genetics, by suggesting a reclassification of humans and the great apes. At the very least, humans, chimpanzees, and gorillas should share the same family status, with orangutans in a separate family. If there is a line to be drawn in this evolutionary scheme of things, it will have to be one that places the African apes in our company, not one that excludes them. Humans are African apes, of an unusual kind. Similarly, if a mental Rubicon was crossed in our evolutionary history, humans are not the sole occupants of one bank; they are accompanied there by apes.

The distinctiveness that we have so assiduously ascribed to ourselves as humans is, in reality, an accident of history. Imagine, for instance, how much more distinct we could have claimed our species to be had all the great apes become extinct before we began pondering our position in the world of nature. If vervet monkeys were our closest relatives, humans would indeed appear to stand separate. Equally, if the species of hominid that links us to our common ancestor with the African apes had not become extinct, the gap between us and chimpanzees would be closed all the way. Gradations between human and ape would be present at every step, and our revered distinctiveness would vanish.

It is simply a contingent fact of history that certain species did become extinct during the past five million years, leaving us to compare ourselves with the African apes as our closest living relatives. And it is a sobering fact of current history that the comparison between humans and apes may soon become virtually artificial, as each species of ape faces extinction in its natural populations. If this happens, it means we will lose the opportunity to learn about ourselves from our nearest living relatives, just at the time that we have indeed recognized them as our relatives. It also means that we will have frittered away our one remaining chance to allow our sibling species to live the way of life for which they, and we, co-evolved across the millennia.

Acceptance of our biological and cognitive intimacy with the great apes has profound consequences for the boundary wall that was erected between humans and the rest of the animal world. For the wall represented more than biological classification or righteous superiority. It was also a moral boundary. On one side—ours—important rights were conferred, rights to freedom and justice; on the other side—theirs—rights to freedom and justice were disallowed. It is, for instance, illegal to perform medical experimentation on a brain-dead human, while such activities are perfectly acceptable on a conscious chimpanzee. If we accede to the logic of what we now know about our self-aware, symbol-using biological relatives, then the boundary must be shifted to include the great apes on our side; and this includes a shift in the moral boundary, too. The moral boundary, artificially erected by us, is no longer defensible.

What if apes were granted something of a semi-human legal status? What if their emotions, intellect, and consciousness were to be widely judged at least morally equivalent to that of children who suffer cognitive impairments? All the data we have gathered over the past twenty years at the Language Research Center while working side by side with such children and with apes increasingly support this view.

What would be the implications of such a view? We certainly would not put these children in a zoo to be gawked at as examples of nature, nor would we permit medical experimentation to be conducted with them. Behavioral experimentation would be permitted only to the degree that it could be expected to enrich and aid their lives. Nor would we put them in a reserve where they could lead a natural life, doing just as they pleased. On the other hand, we can hardly argue for mainstreaming apes as is currently popular to press for with such children. Apes would not quite fit in our society, as their physical prowess is far beyond that of our own.

The currently fashionable answer is to leave them alone on

reserves in the wild. At one time, this seemed like an ideal solution for Native Americans as well. Even if it works on a temporary basis, it does not address the problem of what to do with all the apes that have currently been brought up in the civilized world and do not now know how to make their way in the wild. It also does not tell us what to do when a reserve becomes overpopulated with apes and what to do if they should decide to wander beyond its boundaries.

The future is full of dilemmas. As each one comes into clearer focus, it is easy to see why man has erected a barrier between himself and the other animals on the planet. This barrier has freed us from responsibilities that we, as a species, were not able to meet. I hope that now we are ready for the challenge, for if we meet it, we shall surely build a better world, one in which man and animals walk side by side with a new understanding, a new respect, and a new recognition that each is but a different physical manifestation of life forces, each seeking to make itself known and to live in harmony with the other.

References

Chapter 1

1. Cited in Stephen Jay Gould, "Bound by the Great Chain," *Natural History*, November 1983, 20–24, p. 24.
2. Roy Chapman Andrews, *Meet Your Ancestors* (New York: John Lang, 1948), 11.
3. Phillips Verner Bradford and Harvey Blume, *Ota Benga: The Pygmy in the Zoo* (New York: St. Martin's Press, 1992), 5.
4. Ibid.
5. Charles Darwin, *The Expression of Emotions in Animals and Man* (London: John Murray, 1871), 7.
6. George J. Romanes, *Animal Intelligence* (London: Kegan Paul, Trench, & Co., 1886), 6.
7. Ibid., 429.
8. Alfred Russel Wallace, *Darwinism* (London: Macmillan, 1889), 469.
9. Ibid., 463.
10. Ibid.
11. Leslie White, *The Science of Culture: A Study of Man and Civilization* (New York: Grove Press, 1949).
12. Matt Cartmill, "Human Uniqueness and Theoretical Content in Paleoanthropology," *International Journal of Primatology* 11 (1990): 173–192, p. 178.
13. Thomas Henry Huxley, *Evidence as to Man's Place in Nature and Other Anthropological Essays* (New York: D. Appleton and Company, 1900), 155–156.

14. Noam Chomsky, *Language and Problems of Knowledge* (Cambridge, MA: M.I.T. Press, 1988), 34.
15. Ibid.
16. William McGrew, "The Intelligent Use of Tools," in *Tools, Language and Cognition in Human Evolution*, eds. Kathleen R. Gibson and Tim Ingold (Cambridge: Cambridge University Press, 1993), 151–170, p. 158.
17. Christophe Boesch, "Aspects of Transmission of Tool-Use in Wild Chimpanzees," in *Tools, Language and Cognition*, eds. Gibson and Ingold, 171–183, p. 177.

Chapter 2

1. Robert Yerkes, *Almost Human* (New York: Century Company, 1925), 180.
2. R. Allen Gardner and Beatrice T. Gardner, "Comparative Psychology and Language Acquisition," *Annals of the New York Academy of Sciences* 309 (1978): 37–76, p. 73.
3. Thomas Sebeok, "Performing Animals," *Psychology Today*, November 1979, 78–91, p. 79.
4. Nicholas Wade, "Does Man Alone Have Language?" *Science* 208 (1980): 1349–1351, p. 1349.
5. Thomas Sebeok and Jean Umiker-Sebeok, cited in Wade, "Does Man Alone Have Language?" 1351.
6. Herbert Terrace, "Why Koko Can't Talk," *The Sciences*, December 1982, 8–10, p. 8.
7. Gardner and Gardner, "Comparative Psychology," 73.
8. Herbert Terrace, "How Nim Chimpsky Changed My Mind," *Psychology Today*, November 1979, 65–76, p. 67.
9. Ibid., 75.
10. Herbert Terrace et al., "Can an Ape Create a Sentence?" *Science* 206 (1979): 892–902, p. 892.
11. Ibid., 901.
12. E. Sue Savage-Rumbaugh and Duane R. Rumbaugh, "A Response to Herbert Terrace's Paper, Linguistic Apes," *The Psychological Record* 30 (1980): 315–318, p. 318.
13. E. Sue Savage-Rumbaugh et al., "Do Apes Use Language?" *American Scientist* 68 (1980): 49–61, p. 61.

Chapter 3

1. Robert Epstein et al., "Symbolic Communication Between Two Pigeons," *Science* 207 (1980): 543–545, p. 545.
2. E. Sue Savage-Rumbaugh and Duane Rumbaugh, "Requisites of Symbolic Communication," *The Psychological Record* 30 (1980): 305–318, p. 305. See also: E. Sue Savage-Rumbaugh. *Ape Language: From Conditioned Response to Symbol.* New York: Columbia University Press, 1986.
3. Terrace, "How Nim Chimpsky Changed My Mind," 67–68.

Chapter 4

1. Yerkes, *Almost Human*, 255.
2. Robert M. Yerkes and Blanche W. Learned, *Chimpanzee Intelligence* (Baltimore: Williams and Wilkins Company, 1925), 48–49.
3. Ibid., 31.
4. Yerkes, *Almost Human*, 248.
5. Dirk Thys van den Audenaerde, "The Tervuren Museum and the Pygmy Chimpanzee," pp. 3–11, in *The Pygmy Chimpanzee: Evolutionary Biology and Behavior*, ed. Randall Susman (New York: Plenum Press, 1984), 3.
6. Harold J. Coolidge, "Historical Remarks Bearing on the Discovery of *Pan paniscus*," pp. ix–xiii, in *The Pygmy Chimpanzee*, ed. Susman, xii.
7. Ibid., xii.
8. Harold J. Coolidge, "*Pan paniscus*: Pigmy Chimpanzee from South of the Congo River," *American Journal of Physical Anthropology* 17, no. 1 (1933): 1–57, p. 56.
9. Adrienne Zihlman et al., "Pygmy Chimpanzee as a Possible Prototype for the Common Ancestor of Humans, Chimpanzees, and Gorillas," *Nature* 275 (1978): 744–746, p. 744.
10. Charles Darwin, *The Descent of Man and Selection in Relation to Sex* (London: John Murray, 1871), 199.
11. Adrienne Zihlman, "Pygmy Chimps, People, and the Pundits," *New Scientist*, 15 November 1984, 39.
12. Henry M. McHenry, "The Common Ancestor," pp. 201–230, in *The Pygmy Chimpanzee*, ed. Susman, 218.

13. Takayoshi Kano, *The Last Ape* (Palo Alto, CA: Stanford University Press, 1992), ix.
14. Ibid., x.
15. E. Sue Savage-Rumbaugh and Beverly J. Wilkerson, "Socio-Sexual Behavior in *Pan paniscus* and *Pan troglodytes*: A Comparative Study," *Journal of Human Evolution* 7 (1978): 327–344, p. 337.
16. Kano, *Last Ape*, 192.
17. Savage-Rumbaugh and Wilkerson, "Socio-Sexual Behavior," 341.
18. Kano, *Last Ape*, 162.
19. Ibid., 209.
20. Ibid., 211–12.
21. Ibid., 91.
22. Ibid., 182.
23. Ibid., 102.
24. Ellen J. Ingmanson, "Branch Dragging by Pygmy Chimpanzees at Wamba, Zaire," *American Journal of Physical Anthropology* 78 (1989): 244.
25. Kano, *Last Ape*, viii.

Chapter 5

1. E. Sue Savage-Rumbaugh et al., "Spontaneous Symbol Acquisition and Communicative Use by Pygmy Chimpanzees (*Pan paniscus*)," *Journal of Experimental Psychology: General* 115, no. 3 (1986): 211–235, p. 214.
2. E. Sue Savage-Rumbaugh et al., "Language Learning in Two Species of Apes," *Neuroscience and Biobehavioral Reviews* 9 (1985): 653–665, p. 653.
3. Savage-Rumbaugh et al., "Spontaneous Symbol Acquisition," 223–224.
4. Darwin, *Descent of Man*, 54.
5. Savage-Rumbaugh et al., "Spontaneous Symbol Acquisition," 214.

Chapter 6

1. Patricia Marks Greenfield and E. Sue Savage-Rumbaugh, "Grammatical Combinations in *Pan paniscus*: Processes of

Learning and Invention in the Evolution and Development of Language," in *"Language" and Intelligence in Monkeys and Apes: Comparative Developmental Perspectives*, eds. Sue T. Parker and Kathleen Gibson (Cambridge: Cambridge University Press, 1990), 545.
2. Ibid., 541.
3. Ibid., 571.
4. Herbert Terrace, cited in "Clever Kanzi," *U.S. News & World Report*, 5 November 1990, 68.
5. Thomas Sebeok, "Chimp Appears to Have Toddler's Grasp of English, *Indianapolis Sun*, 7 April 1991.
6. Noam Chomsky, "Clever Kanzi," *Discover*, March 1991, 20.
7. Elizabeth Bates, Commentary in "Language Comprehension in Ape and Child," ed. E. Sue Savage-Rumbaugh, *Monographs of the Society for Research in Child Development*, 58, nos. 3–4 (1993), University of Chicago Press, 222–242, p. 240.
8. "Language Comprehension," ed. Savage-Rumbaugh, 79–80.
9. Bates, Foreword to "Language Comprehension," ed. Savage-Rumbaugh, 240.

Chapter 7

1. Duane M. Rumbaugh, Interview with authors, 5 February 1993.
2. Duane M. Rumbaugh et al., "The LANA Project: Origin and Tactics," in *Language Learning by a Chimpanzee*, ed. Duane M. Rumbaugh (New York: Academic Press, 1977), 87–90, p. 88.
3. D. Guess et al., "Children with Limited Language," in *Language Intervention Strategies*, ed. R. L. Schiefelbusch (Baltimore: University Park Press, 1978), 101–143, p. 105.
4. Rumbaugh, Interview with authors.
5. Mary Ann Romski et al. "Establishment of Symbolic Communication in Persons with Severe Retardation," *Journal of Speech and Hearing Disorders* 53 (1988): 94–107, p. 94.
6. Ibid., 94.
7. Mary Ann Romski, cited in Michael Richter, "The Origins of Language," *Profile*, September 1984, 1–5, p. 5.
8. Romski et al., "Establishment of Symbolic Communication," 103–104.
9. Rose Sevcik, cited in Richter, "Origins of Language," 5.
10. Adele A. Abrahamsen et al., "Concomitants of Success in

Acquiring an Augmentative Communication System," *American Journal on Mental Retardation*, 93, no. 5 (1989): 475–496, p. 489.

11. Rumbaugh, Interview with authors.
12. Mary Ann Romski and E. Sue Savage-Rumbaugh, "Implications for Language Intervention Research: A Nonhuman Primate Model," pp. 355–374, in Savage-Rumbaugh, *Ape Language*, 358.
13. Mary Ann Romski, Interview with authors, 4 February 1993.
14. Rose Sevcik, Interview with authors, 5 February 1993.
15. Ibid.
16. Mary Ann Romski and Rose A. Sevcik, "Language Learning through Augmented Means," pp. 85–104, in *Enhancing Children's Communication: Research Foundations for Intervention*, eds. Ann P. Kaiser and David B. Gray (Baltimore: Paul H. Brooks, 1993), 90.
17. Ibid., 91–92.
18. Sevcik, Interview with authors.
19. Romski and Sevcik, "Language Learning," 94.
20. Duane M. Rumbaugh, Interview with authors, 5 February 1993.

Chapter 8

1. Kenneth P. Oakley, *Man the Tool-Maker*, 6th ed. (London: British Museum [Natural History], 1972), 3.
2. Darwin, *Descent of Man*, 144.
3. Sherwood Washburn and C. S. Lancaster, "The Evolution of Hunting," pp. 293–303, in *Man the Hunter*, eds. Richard B. Lee and Irven DeVore (Chicago: Aldine, 1968), 293.
4. Thomas Wynn and William C. McGrew, "An Ape's View of the Oldowan," *Man (N.S.)*, 24 (1989): 383–398, p. 383.
5. Nicholas Toth, "The First Technology," *Scientific American* 255, no. 4 (April 1987): 112–121, p. 117.
6. Wynn and McGrew, "Ape's View of the Oldowan," 387.
7. Ibid., 388.
8. Ibid., 389.
9. Ibid., 394.
10. Nicholas Toth, Interview with authors, 15 May 1993.
11. Ibid.

12. Ibid.
13. Adapted from Nicholas Toth et al., "Pan the Tool-Maker: Investigations into the Stone Tool-Making and Tool Using Capabilities of a Bonobo (*Pan paniscus*)," *Journal of Archeological Science* 20 (1993): 81–91, p. 89.
14. Nicholas Toth, "Archeological Evidence for Preferential Right-Handedness in the Lower Pleistocene, and Its Possible Implications," *Journal of Human Evolution* 14 (1985): 607–614, p. 612.
15. Toth, "The First Technology," 117.
16. Toth, Interview with authors.
17. Toth et al., "Pan the Tool-Maker," 89.
18. Toth, Interview with authors.

Chapter 9

1. Dean Falk, Commentary to a paper, *Current Anthropology* 30, 141–142, p. 142.
2. Terrence W. Deacon, "Brain-Language Coevolution," pp. 1–35, in *The Evolution of Human Languages*, eds. J. A. Hawkins and Murray Gell-Mann (Reading, MA: Addison-Wesley, 1990), 4.
3. Edmund S. Crelin, *The Human Vocal Tract: Anatomy, Function, Development, and Evolution* (New York: Vantage Press, 1987), 87.
4. Patricia Marks Greenfield, "Language, Tools, and Brain," (Paper for Wenner-Gren conference, "Tools, Language, and Intelligence," 16–24 March, Cascais, Portugal), 28, 30.
5. Glynn L. Isaac, "Stages of Cultural Elaboration in the Pleistocene," in *Origins and Evolution of Language and Speech*, *Annals of the New York Academy of Sciences* 280 (1976): 275–288, p. 283.
6. Wynn and McGrew, "Ape's View of the Oldowan," 394.
7. Nicholas Toth and Kathy Schick, "Early Stone Industries and Inferences Regarding Language and Cognition," pp. 346–362, in *Tools, Language, and Cognition*, eds. Gibson and Ingold, 350.
8. Ibid., 351.
9. Thomas Wynn, "Two Developments in the Mind of Early *Homo*," *Journal of Anthropological Archaeology*, in press.
10. Ibid.

11. Iain Davidson and William Noble, "Tools and Language in Human Evolution," pp. 363–388, in *Tools, Language, and Cognition*, eds. Gibson and Ingold, 363.
12. Iain Davidson, Interview with authors, 13 April 1993.
13. Davidson and Noble, "Tools and Language," 367.
14. Davidson, Interview with authors.
15. Toth, Interview with authors.
16. Iain Davidson and William Noble, "The Archaeology of Depiction," *Current Anthropology* 30 (1989): 125–155, p. 132.
17. Ibid., 134.
18. Charles Darwin, *The Descent of Man and Selection in Relation to Sex*, 2d. ed. (London: John Murray, 1874), 66.

Chapter 10

1. Cartmill, "Human Uniqueness," 187.
2. Mary Midgely, *Animals and Why They Matter* (Atlanta: University of Georgia Press, 1985), 56.
3. Andrew Whiten and Richard W. Byrne, cited in Roger Lewin, "Do Animals Read Minds?" *Science* 238 (1987): 1350–1351, p. 1350.
4. Richard W. Byrne and Andrew Whiten, "The Thinking Primate's Guide to Deception," *New Scientist*, 3 December 1987, 54–57, p. 54.
5. Dorothy Cheney and Robert Seyfarth, *How Monkeys See the World* (Chicago: University of Chicago Press, 1990), 214.
6. Robert Seyfarth and Dorothy Cheney, "Inside the Mind of a Monkey," *New Scientist*, 4 January 1992, 25–29, p. 29.

Index